Paleozoic Fossils

Bruce L. Stinchcomb

Schiffer Publishing Ltd

4880 Lower Valley Road Atglen, Pennsylvania 19310

Acknowledgements

First, I wish to acknowledge assistance from Patricia Eicks and Mike Fix whose comments and proofreading were greatly appreciated. In addition I would like to acknowledge the following who, in various ways, direct or indirect, encouraged the initiation, progress, and completion of this work and its predecessor *The Worlds Earliest Fossils*. Henry N. Andrews, Washington University; Nick Angeli, Chris Braught, Tom Bevenridge, Missouri Geological Survey; Don L. Frizzel, University of Missouri-Rolla; Dorothy Echols, Washington University; Wallace Howe, Missouri Geological Survey; James Houser, St. Louis Museum of Natural History; Pete Kellams, St. Louis Community College; Adam Marty, Potosi Missouri; Larry Nuelle, Missouri Geological Survey and Brass Rooster Minerals; Alfred C. Spreng, University of Missouri-Rolla; Jim Stitt, University of Missouri-Columbia; Charles Tucker, St. Louis Community College-Meramec; Warren Wagner, Florissant Valley Community College; Robert ("Doc") Watson, Ferguson High School; and Ellis Yochelson, U.S. National Museum and U.S. Geological Survey.

Other Schiffer Books by Bruce Stinchcomb

World's Oldest Fossils, 978-0-7643-2697-4, $29.95

Copyright © 2008 by Bruce L. Stinchcomb
Library of Congress Control Number: 2007940913

Designed by Martha Tyzenhouse
Type set in Impact / New Baskerville BT

ISBN: 978-0-7643-2917-3
Printed in China

Schiffer Books are available at special discounts for bulk purchases for sales promotions or premiums. Special editions, including personalized covers, corporate imprints, and excerpts can be created in large quantities for special needs. For more information contact the publisher:

Published by Schiffer Publishing Ltd.
4880 Lower Valley Road
Atglen, PA 19310
Phone: (610) 593-1777; Fax: (610) 593-2002
E-mail: Info@schifferbooks.com

For the largest selection of fine reference books on this and related subjects, please visit our web site at
www.schifferbooks.com
We are always looking for people to write books on new and related subjects. If you have an idea for a book please contact us at the above address.

This book may be purchased from the publisher.
Include $3.95 for shipping.
Please try your bookstore first.
You may write for a free catalog.

In Europe, Schiffer books are distributed by
Bushwood Books

Contents

Introduction

This is the second Schiffer book on fossils with a focus on collecting. The first work, *The Worlds Oldest Fossils,* dwelt primarily with the fossil record of the Precambrian. This earliest fossil record by some accounts, perhaps somewhat overstated, has been said to be "just so much algae." This early record of life however, when compared with the rich fossil record of the Paleozoic shows some truth in that statement; the megafossil record of the Precambrian pales when compared to the richness of the Paleozoic Era.

As with *The Worlds Oldest Fossils* there is a certain inherent bias in the work (as there is in all comprehensive works on fossils), a bias toward including those specimens that are available and accessible to the collector and those that have been or are available on the fossil market. There is also a certain bias toward fossils of North America, particularly those from the United States (U.S.) midwest.

The richness of the fossil record of the Paleozoic Era meant that a selection process had to be persued to determine what was to be included in a work of this sort, and to ensure a balance between animals (vertebrates and invertebrates), protists, and plants. The author has also attempted to achieve a balance between those specimens that may be obtained by collecting or exchange and those that have become available on the fossil market, particularly during the last two decades. The book is not a textbook, but it is quite rich in paleontological information relevant to the particular specimen being illustrated and discussed. Paleontology as an avocation has great potential for delving into science and deepening one's understanding of it, involving not only a familiarity with the geologic past, but an understanding of biology and geology as well.

Chapter One
Fossils, Strata, and Geologic Time

A look at fossils.

If you have picked up this book and are reading it, it is probably because of an interest in or curiosity about fossils. Even a cursory browsing through these pages makes it obvious that the study of the fossil record is both broad and deep. No field of serious collecting gets a person involved more with the basic sciences than does paleontology. It is rooted in biology, with its classification systems (taxonomy), evolutionary theory, and related subjects. Geology plays a role, with its megatime frame and the geologic relationships with rocks and rock structures. Some chemistry, biochemistry and physics are also involved.

While this book is written for the fossil collector, almost everyone will find the diverse world of Paleozoic life and its fossils of interest in itself. Fossils are not rare and a casual encounter with them will often arouse an interest in learning more about them, which this book hopes to reinforce. Fossils are a part of and are connected with the rock strata of the earth's crust. Fossils can also be a direct and personal link with that abstraction known as geological or deep time.

Usually found in sedimentary rocks, whose layers can sometimes be filled with them, fossils have attracted the attention of people for centuries. Their occurrence in sedimentary rock layers means that fossils are not just a surface phenomena, but extend with the layers in which they are found into the earth's crust itself, often in prodigious numbers.

The study of fossils is called paleontology. This is usually considered as a branch of geology although it is equally home as a part of the life sciences.

Many persons confuse paleontology with archeology. Archeology is the study of the unwritten record of human life; as such it can extend back in time a few million years at most. Paleontology on the other hand, encompasses most of the record of the earth's geologic past, a span of almost four billion years. The two fields utilize different methods of study, particularly in the field. Archeology is site restricted, its resources often being very localized. Fossils occur in rock layers that extend into the earth's crust and can occur over large areas. Paleontology is dependent upon outcrops which, because of excavations, quarrying, and just plain erosion, can yield "new" specimens.

The confusion of paleontology with archeology leads some to apply the more "limited resource" mindset of archeology to paleontology, so they are appalled with encouragement of paleontological collecting. This archeological model has also been used by some governmental agencies and land managers to restrict or prohibit paleontological collecting. They believe they are protecting a valuable scientific resource. Ironi-

Fig. 01-01. Fossil brachiopods can cover limestone beds and can even compose them resulting in prodigious numbers of fossils considering that these beds go into the earth's crust. These are low quality fossil-covered slabs with little or no value which can be found in many areas of the world.

Fig. 01-02. Arrowheads! These stone tools, found in the St. Louis Missouri region were made by native Americans about 1,000 years ago. Such artifacts form part of the resources of archeology which many persons confuse with paleontology. Most arrowheads found in North America are not fossils, they are not old enough.

cally this mindset often results in destruction and loss of more fossils and relevant information than does any amount of collecting, particularly when collectors form a community that works with paleontological "professionals," an arrangement which has had a long standing tradition. Its been said that the more eyes looking for fossils, the more will be found and some of these will have scientific importance.

In a time when its seems more and more persons are scientifically illiterate (particularly in the U.S.), fossil collecting and paleontology offer a way to learn some basic science in an interesting and enjoyable manner. Unfortunately, in the U.S. at least, collecting fossils in the field is no longer as easy as it once was.

As the country has become more structured and populated, the collection of fossils has become more and more difficult and restricted. This is in contrast to many parts of the developing world that provide some striking, high quality fossils for what in the U.S. are low prices. Encouraged by the lack of restrictions, these countries are the source for many of the nice fossils on the market today

But for collection and display, one seeks specimens that have one or more of the following attributes:

(1) they are complete

(2) they are rare, and

(3) they are pleasing to the eye (paleontological eye candy)

Fossils are common in many sedimentary rocks! The relative abundance of many fossils is the reason that many of them have little or no monetary value. With

Fig. 01-03. Establishment of strict trespass laws and concerns with landowner liability in recent years have made it more difficult to collect and enjoy natural history and its objects in many parts of the U.S. This posted creek outcrop was the source of many specimens like those shown in chapter eight.

common fossils, the goal of collecting is that of learning rather than any monetary considerations. Even with rarer complete fossils, when their value goes up, additional field collecting and even fossil mining can produce additional specimens, consequently driving down the price. In many ways the fossil market works like the gem market. Precious gems are usually not really good investments because as their value increases, more gem bearing rock is mined and processed.

What's great about fossils is that they are more readily available and within the reach of the average person than are many gems. All around the world

Fig. 01-04. These trilobites were collected by Moroccan children from barren shale surfaces near where they live.

fossils are waiting to be found by observant fossil hunters, sometimes good ones. Many sedimentary rocks, particularly the geologically younger ones, are full of them. If a person really wants quality specimens (which can have monetary value), they are there also. To do this however, one has to become knowledgeable, focused, and persistent in one's fossil acquisition. Fossil collecting is a democratic activity. They are abundant enough that most persons can find them, if they persevere and look hard enough.

To find really good specimens takes a lot of persistence as nature herself puts up some pretty tough barriers, particularly with desirable fossils. This democratic nature of fossil collecting is one of the reasons that governmental roadblocks to their collecting and acquisition can be onerous.

Fig. 01-05. These fossil crinoids were found by the author as a child in a neighbor's patio slabs. When found they were subtile and only strong interest and perseverance, ranging from replacement of the slabs to learning how to expose the fossils with hand tools, made them desirable specimens.

A Second Look at Fossils and Paleontology.

A cursory look through this book will yield some significant insight into life of the earth's past. For instance, only invertebrates will be found in the Cambrian (chapter 2) in contrast to the Permian Period (chapter 9) when reptiles and amphibians are both present.

There is inevitably going to be some bias in a presentation such as this. There is, for instance, a focus on North American and particularly midwestern U.S. fossils as these were more readily available to the author. There is also a preservational bias for more complete specimens, a bias often found in other paleontological works as well. Thick shelled organism such as mollusks and corals are more likely to be preserved as fossils and to be found than are those with delicate exoskeletons. This results in those organisms which lack hard parts altogether being represented even less; scientifically it is these specimens that are usually the most significant and desirable. Such a preservational bias has been compensated for by the inclusion of fossils from what are known as "fossil lagerstatten," which are fossils from localities where the soft fleshy parts of organisms have been preserved. Some of these localities are the Mazon Creek and Essex ironstone nodules of northern Illinois, the Hunsruck or Bundenbach slates of Germany, as well as other exceptional fossils from localities that are called, appropriately, "paleontological windows." Paleontological "windows" are so-called because they allow an unbiased look at a complete fauna rather than just those organisms with more durable hard parts.

There is also a bias in the work toward specimens available to collectors. Mazon Creek fossils have been widely distributed among collectors, and average and good specimens are available and accessible. Burgess Shale fossils and some vertebrates, by contrast, have been prohibited from non-institutional ownership and

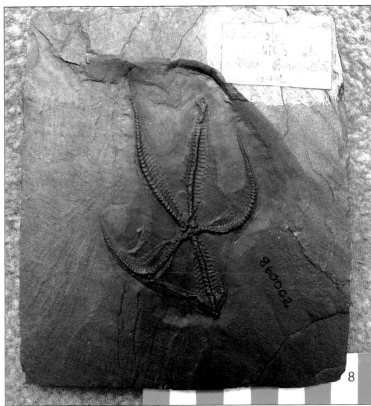

Fig. 01-06. Soft bodied preservation and complete fossils are most desirable. This starfish from one of the paleontological windows, the Devonian Bundenbach slates of Germany, shows the preserved soft body tissue around its arms.

are not represented here for that reason. Fossils from other localities are also not included as specimens are rarely seen as a consequence of uncooperative land managers and/or owners.

Because the world of Paleozoic fossils is vast, this work should not be looked upon as being exhaustive. Its predecessor, *The World's Oldest Fossils*, attempted to cover Precambrian megafossils with an element of completeness and with these early fossils that could be accomplished to some degree. There was however such a diversification of life in the Paleozoic Era that any attempt at completeness might well be considered audacious.

In assembling this book, the aim was to present a mix of quality common fossils and museum quality specimens. In many regions of Paleozoic rocks, particularly on the craton, numerous brachiopods, corals, bryozoans, and mollusks can be found in great abundance. If one is trying to understand and identify fossils which one has collected locally, regional geologic literature (state, provincial and museum publications) are usually the best source of information when such are available. What is presented here are those fossil groups and specimens which are either significant or are aesthetically pleasing. Beyond their scientific interest, fossils can be, after all, one of natures "objects de art." Rare fossils are considered, both those that are rare because they are found in limited numbers and those whose rarity is in their completeness. It should be emphasized that with rare fossils, which may have relatively high monetary value, knowledgeable and thoughtful collectors always consider "that science comes first."

Some Sources of Fossils for a Collection

Besides the collecting of specimens oneself, an activity requiring familiarity with local geology (and an excellent way to get to know it), fossil specimens are available from a number of sources. One source is roadside rock shops. Rock shops in the U.S. however, seem to becoming a thing of the past.

The author recalls rock shops from the 1950s through the 1990s having locally collected fossils and minerals, often at low prices as these were usually obtained at little or no cost. Such rock shops have been supplanted by local gem, mineral, and fossil fairs in most larger urban areas. Two fossil related fairs that occur in the U.S. are particularly noteworthy: the Tucson, Arizona, show in late January and the MAPS Expo (Mid American Paleontological Society), a strictly fossil-oriented show. The latter is usually held in late March or early April in Macomb, Illinois. The Tucson show is the largest such activity in the world, held in several locations around the city. Fossil, mineral, and

gem dealers from many parts of the world display and sell their wares. The scale of the Tucson show is such that it is nearly impossible for one person to take it all in, even with specific areas being devoted to minerals and fossils. The MAPS expo is more manageable, with a considerable variety of fossils, many at wholesale prices, being spread out in a large gymnasium. The third major show in the U.S. is the Denver, Colorado, Show in September. This is somewhat like the Tucson show in its diversity (minerals, gems and fossils), but is less spread out. The other source of specimens is the internet. Quality fossil specimens (and some not so great) show up regularly on the net at places like E-bay. The net appears to be enlarging interest in fossils on a world wide basis.

Fig. 01-07. Fossils from a 1950s roadside rock shop in southwestern Missouri along U.S. Route 66. These brachiopods were collected locally, from chert boulders on rocky hillsides of the western Ozarks

Geologic Time Scale

The development of the geological time scale began early in the 19th century. It started with the realizations that fossils were the remains and evidence of life from the geologic past, and that there was, in fact a geologic past, that the world was inconceivably old. Concepts that in the 18th century were hazy and unclear, took on a new clarity and significance.

Nineteenth century explorations led to a growing number of classifications and definitions. When following rock strata downward (and hence backward in geologic time) geologists encountered a point below which fossils appeared to be absent. Using this as a starting point, the fossil-yielding strata just above these unfossiliferous *primordial* layers were at first called

Silurian, after a place in Wales where such strata were first observed. After some convoluted arguments, this layer was named after Cambria in Wales and became known as the Cambrian System. The Cambrian Period of geologic time is that part of geologic time whose rocks belong to the Cambrian System of strata. Around the world, Cambrian age strata have been found to bear the first appearance of abundant fossils. Older and seemingly unfossiliferous strata became known as Pre-Cambrian strata, referring to the fact that such strata was formed previous to the Cambrian.

For a few decades of the mid-19th century, geologists argued that the Cambrian, with its seemingly sudden appearance of fossils, marked the creation of life; that is that time in earth history when God said "Let there be life, and there was life!" Charles Darwin, in his "Origin of Species", discussed this geologic dilemma by stating "The case at present must remain inexplicable; and may be truly urged as a valid argument against the views (evolution) here entertained." More recently, this sud-den (and still mysterious) appearance of abundant and diverse organisms in the fossil record has become known as the "Cambrian radiation event" and its appearance in the earth's sedimentary rocks marks the beginning of the Paleozoic Era, the subject of this book.

The Paleozoic Era is represented by many life forms that would be strange to us if they lived today. Trilobites, straight cephalopods, blastoids, and bony armor fish, to name a few, were at various times domi-nant life forms of the Paleozoic Era but are now extinct. In fact, most life forms that lived during the Paleozoic Era are now extinct with a few exceptions. Among those Paleozoic organisms that survived, such as brachiopods and crinoids, some were diverse and abundant during the Paleozoic Era, but today are greatly reduced in both their diversity and distribution.

The end of the Paleozoic Era saw the disappear-ance of most of the life forms presented in this book as well as many others. Distinctive animals such as trilobites and straight cephalopods characterize the Paleozoic Era and these disappeared rather abruptly at the end of the Permian, the last geologic period of the Paleozoic. This mass extinction, the most profound extinction in the history of life, is known as the termi-nal Paleozoic extinction event.

Order of Images

Chapters are arranged in chronological order in which they occur in the earth's crust, with each chapter representing a geologic period of the Paleozoic era, with the Cambrian occurring first. The Lower Ordovi-cian (Ozarkian-Canadian or Ibexian) is second, as its life was distinctively different from that of the Middle and Upper parts of the Ordovician Period which make up the third chapter. The Middle and Upper Ordovi-cian are followed by the Silurian and Devonian periods where each period is represented by a chapter with a focus on those fossils often available through collect-ing and/or fossil fairs or purchase. The Carboniferous is divided into the Mississippian and Pennsylvanian periods (each with a chapter) as has been tradition-ally done in North America. The last chapter of the book deals with fossils of the Permian Period, where we catch a glimpse of the Mesozoic Era and its life forms, particularly the vertebrates.

Within each chapter the plants and protists are presented first, followed by animals. In the animal kingdom the presentation is from lower invertebrates (sponges, cnidarians, bryozoans, brachiopods, and annelids to higher invertebrates (echinoderms, mol-lusks, and arthropods). Representatives of the phylum chordata (mostly vertebrates) end each chapter.

Fig. 01-08. This strata is from the "bottom of the stack" of sedimentary rocks containing animal fossils. Below it similar layers essentially lack fossils. This strata marks the beginning of the Cambrian Period. Chilthowee Series, Montevale Springs, eastern Tennessee.

Fossils and Continental Drift

The concept of continental drift was originally proposed in the 1910s by Alfred A. Waegner, a German earth scientist. Continental drift was embraced by some geologists when it was introduced, but others scorned the concept. The same was true among paleontologists. Identical geology and fossils on separate continents, such as those that are common between the Canadian Maritimes and Wales on the other side of the Atlantic, were noted as early as the 1840s by Charles Lyell. Through the decades of the 19th and 20th centuries other earth scientists noted the similarity of rocks, geologic structures, and fossils between adjacent continents. The exotic nature of continental drift theory precluded many earth scientists from accepting the concept. In science "extraordinary claims require extraordinary proofs" and the matching of fossils and rock strata just didn't seem "scientific" enough to muster acceptance.

In the 1960s geophysics, using paleomagnetism measurements and plotting paleomagnetic vectors, provided the necessary extraordinary proof and pretty well proved the existence of continental drift. From this the concept of plate tectonics quickly evolved with its phenomena of sea floor spreading. Sea floor spreading explained another very puzzling dilemma in paleontology: the presence, in some regions, of two separate sets of strata of the same geologic age, each bearing a different set of fossils. Such autochronous terrain was originally explained by the presence of barrier islands, or some other type of marine barrier which prevented two nearby faunas from intermixing. The concept of sea floor spreading, with its ability to transport parts of the ocean floor and its sediments thousands of miles and attach them to a continental margin, finally solved this paleontological dilemma. The two series of rock strata and their similar age faunas originally came from entirely different parts of the globe.

Geology has always recognized that there were a number of different environments under which sediments could be deposited and that these were reflected in the associated fossils. Plate tectonics enabled scientists to relate various sedimentary environments with specific tectonic regions of the earth.

Deep Sea Sediments

The origins of sedimentary rock of a type known as flysch was always puzzling. Flysch deposits, obviously deposited on the ocean floor, lack fossil shells or calcareous hard parts but sometimes contain abundant trace fossils (tracks and trails of animals).

Plate tectonics explained flysch deposits as representing the sediment of the deep sea. Sediment deposited in deep bodies of water and then transported by sea floor spreading became folded and compressed when they encountered a continent and are now often found forming parts of the rocks of mountain ranges.

Fig. 01-09. These tectonically deformed beds were originally deposited on the floor of the ocean. Such deep sea sediments were later transported, in conveyer belt fashion, by sea floor spreading and deposited at the edge of the North American continent.

Geologists in the late 19th century recognized that many of the earth's mountain ranges were formerly the sites of basins, often a subsiding basin with sediment filling it to produce thick sequences of sedimentary rock. Such thick sequences of sedimentary rock became known as geosynclinal sequences and the basin or trench in which these great thicknesses of sediment were deposited is called a geosyncline. Two main types of geosynclines exist, one containing

Fig. 01-10. Limestone beds deposited in a geosyncline near the edge of the North American continent were later compressed and folded to form parts of the Appalachian Mountains.

some limestone beds and often rich in marine fossils with calcareous shells, the other usually barren of such marine shells but containing trace fossils and, in the Lower Paleozoic, peculiar fossils found in black, slaty shale known as graptolites. These two types of geosynclines were termed miogeosyncline and eugeosyncline respectively.

Flysch deposits, with their trace fossils and lack of fossil shells, were in the later category. Plate tectonics later showed that miogeosynclines represented sediment deposited at the margin of continents in a gradually sloping and subsiding "ramp" where the water depth became deeper nearer the edge of the continent. Eugeosynclines on the other hand, repre-

Fig. 01-11. Fossil marine animal burrows (trace fossils) in a large slab of Lower Paleozoic sandstone from strata deposited near the edge of a geosyncline near what is now Logan, Utah.

Fig. 01-13. Geosynclinal sediments west of those in the previous picture were formed near the edge of what, in Cambrian Period, was North American continent. Antelope Springs, House Range, western Utah.

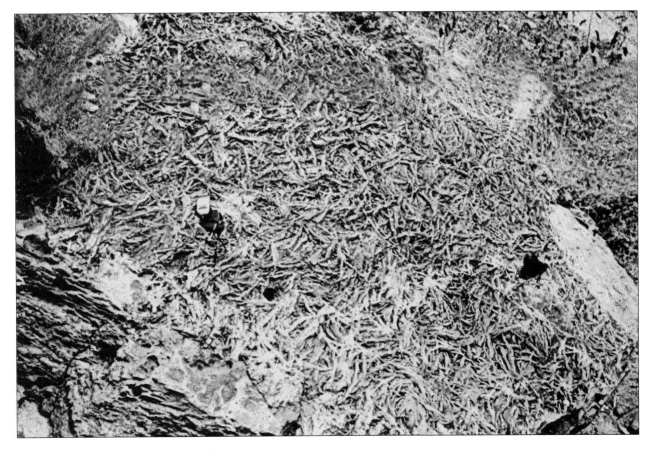

Fig. 01-12. Close-up of the trace fossils in the previous picture.

sent either the deposits of the open ocean or of deep sea trenches which the process of sea floor spreading moved and later piled onto the edge of a continent. Eugeosynclines essentially represent sediment scraped off of the sea floor and deposited or piled against the continent by the process of sea floor spreading.

Rift Zone Sediments

The opening of rift zones and the consequent formation of a "new" ocean basin explains another type of sedimentary domain and its associated fossils. This can consist of thick masses of dirty sandstone (greywacke) or of arkosic sandstone, both lacking marine fossils.

The plant and fresh water fish bearing sediments (terrestrial sediments) of the northern Appalachians are examples of sediments deposited in this environment (See Chapter 6 on the Devonian). As the rift zone opens more sediment is introduced into it. Such sediment has a distinctive signature about it, partially as it is often made of terrestrial rather than marine sediments.

Fig. 01-14. Geosynclinal sediments in central Alabama of Cambrian age (Rome Formation) were deposited at what, at the time, was the edge of the North American continent.

Fig. 01-16. Bluff of Pennsylvanian sediments deposited in a rift zone where the sediments have been compressed and almost tilted on end. Joggins, Nova Scotia.

Fig. 01-17. Another sequence of sediments near Joggins, Nova Scotia, deposited in a rift zone.

The Craton

Sedimentary rocks of both the deep sea (Eugeosyncline) and rift zones are associated with intense tectonic activity (earth movements). Such sedimentary rocks are generally folded and/or tilted in their outcrops and in fact are rarely found in the horizontal position in which they were originally deposited. This contrasts with sediments deposited on a stable, continental crust, where such sediments may be relatively thin and lying upon a "basement" composed of very ancient and hard rock like granite.

Fig. 01-15. Rift zone sediments of the Upper Devonian in which fish and land plants are preserved. Escuminac Formation, Escuminac, Quebec

Sediments deposited on a stable continental surface, or on what is known as the craton, will usually be found to be in the horizontal position in which they were deposited, even though such sediments may be hundreds of millions of years old. This is in stark contrast to those sediments from the ocean floor or those deposited in rift zones, which will be tilted, folded, or crumpled. Underlying sediments of the craton are sequences of very old rocks, often indicative of very ancient tectonic activity but later incorporated into part of a continent where there was then much tectonic stability. Horizontal sedimentary rocks, which are often rich in limestone, represent those sediments spilling onto the stable continent when it was depressed by tectonic activity in some adjacent area but not severely dragged down as are the sediments associated with an active tectonic area. This spilling or depositing of marine sediments along a continental margin dragged down by the process of subduction represents shallow water usually with limestone beds potentially full of fossils.

Lake and Lacustrine Deposits

Sedimentary rocks can be deposited in ancient lakes and associated fossils will then be types such as fresh water fish and lake associated insects and plants. Such ancient lake deposits can be deposited both on the craton, in which case they will be horizontal, though in some cases they can be associated with regions of tectonic activity and be tilted, folded, or crumpled.

Fig. 01-18. Dipping (or tilted) strata of Lower Devonian age, Gaspe Peninsula, Quebec. Such strata was originally deposited in a rift zone. The sediment (or sedimentary rock) consists of dirty sandstone and shale beds which contain the early and primitive land plants seen Fig. 06-01.

Fig. 01-20. Horizontal Late Paleozoic sedimentary rocks on the craton. San Juan River, southern Utah.

Fig. 01-19. Horizontal sedimentary rocks of Middle Ordovician age south of St. Louis, Missouri, were deposited on continental crust (craton) and remained undeformed for over 430 million years. Horizontal sedimentary rocks like this are characteristic of sedimentary rocks on the craton

Fig. 01-21. These shales were deposited in a lake which existed during the Mississippian Period in what is now New Brunswick, Canada. They contain significant amounts of petroleum as well as fossil fish (see Figs. 07-97 to 07-99). Albert Mines, New Brunswick, Canada.

The Value of Fossils and Value Range

To the uninformed, the monetary values placed upon fossils that, with a little searching, they could find for themselves, makes fossil hunting seem an easy source of money. Nothing could be further from the truth. In the right areas one can often find fossils, but this is not like "finding money." The value of a fossil depends on type of fossil, its completeness, rarity, and appearance. Fossils can be common; some whole layers of rock are composed of them, but generally these have no monetary value. And even when a fossil is found, it is not ready for the marketplace. Fossils usually don't look like much in the field; they have to be brought out from the rock. Preparation involves the removal of excess rock and the exposure of the fossil itself, often by painstaking manual methods or by the use of specialized tools and technology like pneumatic chisels, air abrasive machines or by the use of acids and other chemicals. The list is extensive so that fossil preparation is a specialty in itself and is more of a art than a science.

The specimens in this book are, for the most part, average to excellent examples from Paleozoic rocks and localities. It would be audacious to attempt the inclusion of all or even most Paleozoic fossils. New material is continuously turning up and new localities are being discovered that result in a "flood" of new specimens of previously known material.

Some have criticized the inclusion of average material and suggesting that the focus should be only on "the very best." Top quality "perfect" specimens are spectacular and beautiful, but they can also be quite pricey and often don't show features of the organism any better than do those less perfect specimens. Such specimens are the "trophy specimens" which are quite desirable, but by restricting a collection to them one's exposure to the great diversity of fossils and paleontology is necessarily limited. Also, if one collects fossils in the field, what will be found will seldom be comparable to top dollar specimens which might be sold by a dealer. Specimens sold by dealers generally have been prepared or cleaned, often with the use of rather expensive equipment like an air abrasive machine or polishing equipment. Fossil preparation is time consuming and along with the fossil's level of perfection and rarity can be cause to "jack up" the price.

The approach here is to illustrate fossils that are available in the marketplace, usually without the outlay of large amounts of cash. This does not mean that quality needs to suffer. Because of availability, some top quality specimens can be relatively inexpensive.

Fossils and land issues

The matter of the monetary value of fossils is a potentially sticky one. There was a time in the U.S. when few persons outside of the geologic community cared about fossils. There was little value placed upon them and there also was a more liberal attitude toward the public's use of private land, a combination that gave considerable freedom to the person who wanted to collect fossils.

Setting a monetary value, on the one hand, must consider the time and skill involved in fossil preparation, which rightly should increase a specimen's value. On the other hand, the fact is that fossils, unlike most other collectables, occur naturally as part of the land and they can be there for the "taking."

This brings up the issue of the ownership of land and its fossils. Private land is one thing; fossils found on private land belong to the land owner. It is the author's opinion, however, that as a "humanistic issue" landowners who have no interest in fossils themselves allow responsible, interested persons access to their land and allow reasonable collecting or at least enable collecting through some sort of reasonable monetary arrangement.

Public land is another matter! The situation regarding collecting on public land (federal, state, and local) today runs from absolute prohibition (it's better that the fossils weather away or be buried than that they be collected) to the liberal attitude of not being overly concerned with the issue. The response to the issue is dependent upon the policies of the particular land managing authority involved and the attitude of its agents. On the federal level in the U.S., this issue was addressed in a 1991 study by the National Academy of Sciences which recommended that non-commercial collecting of most fossils be allowed on most federal lands. Such a policy has been maintained by the U.S. Forest Service and the BLM (Bureau of Land Management). The author is of the opinion that the current policy of these two land management agencies represents a reasonable solution to this problem. With this in mind, it is the policy of the author not to place monetary values on fossils recently collected from public land localities.

Many public land localities, however, such as those of the House Range of Utah, have been extensively collected for many decades and fossils from them frequently turn up in old collections. Monetary value on such specimens has been established and it is believed by the author to be appropriate to report them here. Technical legal aspects of fossil, mineral, meteorite and rock collecting can really be a "can of worms," so that often the best approach is to use common sense in your collecting.

The value ranges used in this book

A $1,000–2,000
B $500–1,000
C $250–500
D $100–250
E $50–100
F $24–50
G $10–25
H $1–10

Bibliography

Seyfert, Carl K. and Leslie Sirkin, 1979. *Earth History and Plate Tectonics*. Harper and Row, New York, London.

Shimer, Hervy W. and Robert Shrock. *Index Fossils of North America*. John Wiley and Sons and/or The Paleontological Society, 1944 (1985).

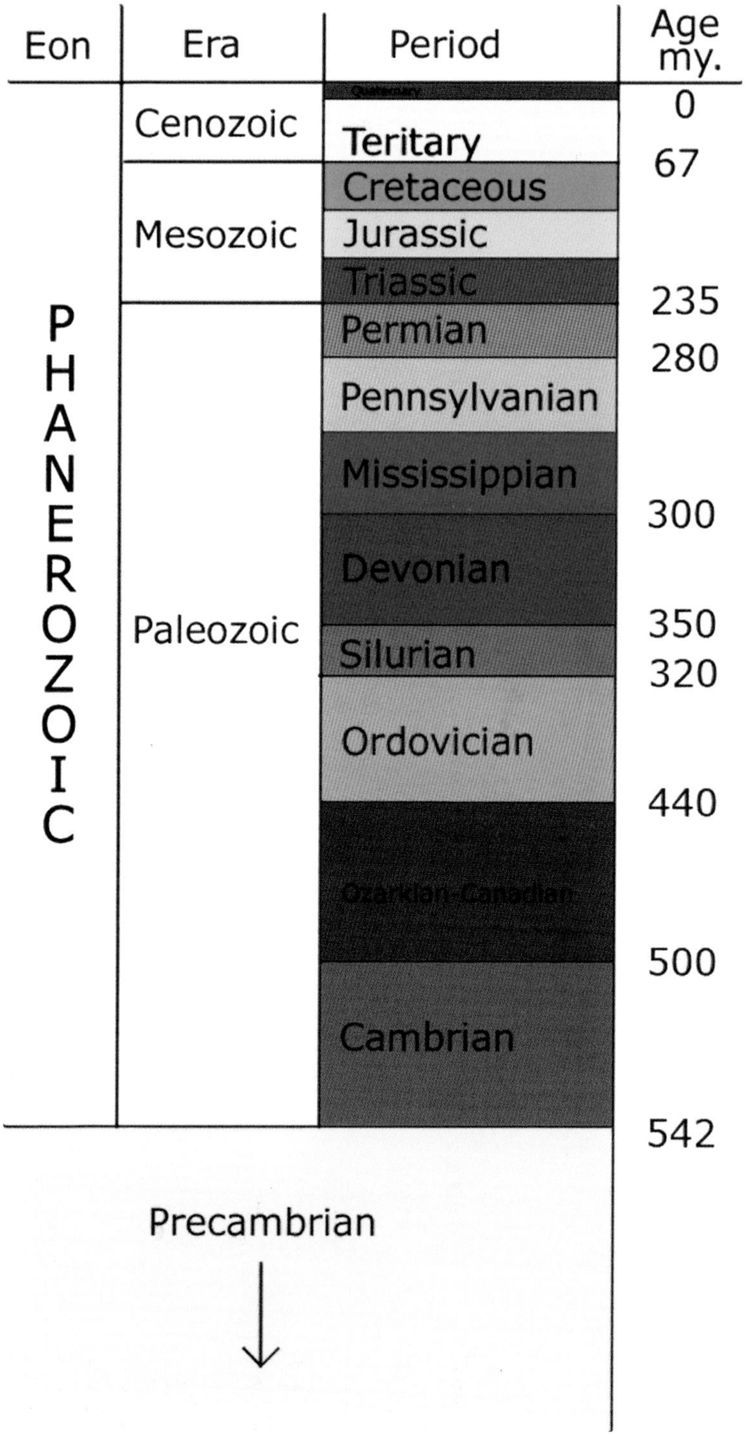

Eon	Era	Period	Age my.
P H A N E R O Z O I C	Cenozoic	Quaternary	0
		Teritary	67
	Mesozoic	Cretaceous	
		Jurassic	
		Triassic	235
	Paleozoic	Permian	280
		Pennsylvanian	
		Mississippian	300
		Devonian	350
		Silurian	320
		Ordovician	440
		Ozarkian-Canadian	500
		Cambrian	542

Precambrian

↓

Fig. 01-22. The Phanerozoic geologic time scale.

Chapter Two
The Cambrian Period

The First Period of the Paleozoic Era

The Cambrian Period saw the (seemingly) sudden appearance of most invertebrate phyla in the fossil record. Many mid-19th century geologists thought the Cambrian marked that time in earth history that saw the creation of life. Charles Darwin, when he wrote his "Origin of Species," was mystified by the phenomena and a satisfactory explanation for this sudden richness in the fossil record is still forthcoming.

Cyanobacteria

These are fossils produced by cyanobacteria (blue-green algae), which can be locally quite abundant in Cambrian age strata. *Girvanella* is a type of oncolite, a biscuit-like nodule produced by the photosynthetic activity of cyanobacteria. Both oncolites and stromatolites are biogenic structures produced by photosynthetic activity of primitive life (cyanobacteria); they are common fossils in the Cambrian. Fossil oncolites and stromatolites go back into the deep time of the Precambrian.

Fig. 02-02. A group of digitate stromatolites on the shallow sea floor. The ecosystem represented here was a very ancient one even in the Cambrian Period, going back over t wo and a half billion years at that time. Digitate stromatolites usually have sediment filling between the "fingers" so that they would be so exposed at any time as shown here. Artwork by Virginia M. Stinchcomb

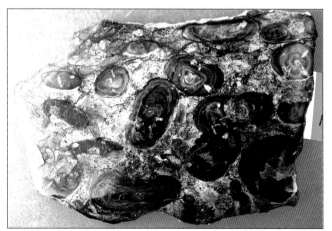

Fig. 02-01. *Girvanella* sp. These elliptically shaped fossils are oncolites. Oncolites are limy globules produced by the physiological activity of cyanobacteria (blue-green algae). The Cambrian period saw colonies of cyanobacteria being a dominant part of ecosystems as they had been for over two billion years before the Cambrian. These are polished slices of relatively large oncolites from the Marble Mountains of southern California. These have been widely distributed among collectors. Poleta Formation, Marble Mountains, California. (Value range G).

Fig. 02-03. *Collenia* sp. This diagonal slice through a colony of digitate stromatolites is hard to distinguish from slices through oncolites. The stromatolites have been cut at 45 degrees so that they resemble slices through oncolites. They come from Upper Cambrian beds forming the front range of the Canadian Rockies of Alberta. White veins cutting the stroms are calcite veins, these were formed by stresses produced during uplift of the Canadian Rockies, the resulting cracks filling with calcite.

Fig. 02-04. *Girvanella* sp. These small oncolites, on examination in thin section, exhibit spaghetti-like fossil algal filaments. They constitute part of a cyanobacterial ecosystem which goes back deep into geologic time. Bonneterre Formation, Ste. Francois County, Missouri. (Value range G).

Fig. 02-06. Land plant? Puzzling, carbonaceous films such as this are found in both the Precambrian and Cambrian and are generally attributed to cyanobacterial colonies such as the gelatinous globules of *Nostoc* sp., a still living, modern cyanobacteria. The carbonaceous compressions represented here are thicker than Precambrian *Nostoc*-like fossils and suggest a more firm, organic composition like one containing cellulose. Cellulose is the organic compound associated with land plants (it's what composes wood), so these compression fossils could have come from some sort of small land plant. If so, they would represent some of the earliest evidence of land plants. It is unclear, however, when the first plant appeared on the land and what type of plant it might have been. These fossils come from a series of "fossil" mud flats around which some type of land plants may have grown. "False" Davis layer of Bonneterre formation, Fredericktown, Missouri. (Value range G).

Fig. 02-05. *Collenia* sp. Digitate stromatolites, like these, formed a series of stromatolite reefs around the Precambrian "core" of the Ozark Uplift of Missouri. Part of this stromatolite reef became the porous rock which hosted heavy metal sulfide mineralization, specifically the lead sulfide (galena) mineralization of the Viburnum Trend, which is part of the lead belt of southeastern Missouri. Such digitate stromatolites represent an algae based ecosystem which goes far back into the geologic past. Stromatolites like these are the dominant fossils through most of the 2.5 billion years of geologic time prior to the Cambrian Period. (Value range F).

18

Archeocyathids

Fossil archeocyathids resemble corals, though they are not. Archeocyathids are considered by some paleontologists to represent an extinct animal phylum; others consider them to be a type of peculiar sponge.

Fig. 02-07. *Cambriocyathus* sp, an archeocyathid. Archeocyathids are strictly Cambrian fossils which have attributes of both corals and sponges. They are found only in the Cambrian and are considered to represent an extinct phylum or a peculiar type of sponge. They were some of the first reef-forming animals, Forteau Formation, Lower Cambrian, Southern Labrador. (Value range G).

Fig. 02-08. A group of small archeocyathids from southern Labrador. (Value range G).

Chancelloria

Chancelloria is a puzzling Cambrian fossil. Some forms, like these specimens from Utah, appear to be the small calcareous plates (spicules) of a sponge. Other occurrences of *Chancelloria* find them covering the surface as sclerites of a soft bodied, weird Cambrian organism.

Fig. 02-09. *Chancelloria* sp. This form of *Chancelloria* is probably a sponge. *Chancelloria* can be represented by small, circular bodies, some of which are found in strata of the Tommotian, the earliest part of the Cambrian which lacks trilobites and was deposited prior to trilobites existence or at least to their existence with a mineralized exoskeleton. Wheeler Formation, House Range, Utah. (Value range E).

Medusoid

These are medusoid (jellyfish-like) fossils particularly characteristic of the Cambrian Period.

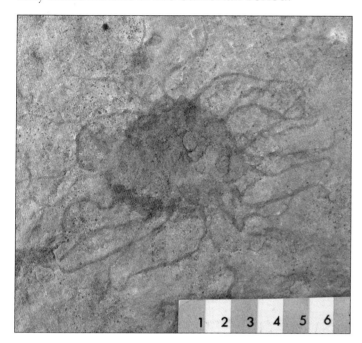

Fig. 02-11. A problematic fossil, suggested to be a medusoid (jellyfish) comes from a trackway and jellyfish-rich series of sandstone beds in central Wisconsin. A number of these have been distributed by a fossil dealer in which the organisms tentacles(?) have been highlighted. Mt. Simon Formation, Middle Cambrian, Mosinee, Wisconsin. (Value range F).

Fig. 02-12. *Dactyloidites asteroides*, Fitch. These fossils were considered as jellyfish impressions by Charles D. Walcott, a late 19th and early 20th century worker on Precambrian and Cambrian paleontology. They have also been considered as trace fossils and as pseudofossils. The trace fossil designation appears to be the most logical as they show up on cleavage surfaces and within beds of slate when the slate is either cleaved or sawed for flooring tiles. Fragile jellyfish impressions would not do this. *Dactyloidites* are found in one horizon in the slate quarries of Middle Granville, New York, which are part of the slate beds that continue into nearby Vermont. Metawee slate, Lower Cambrian, Middle Granville, New York. (Value range E).

Opposite
Fig. 02-10. *Chancelloria?* Forms like this radially symmetrical fossil are found in Cambrian strata underneath (hence older) the Cambrian strata the yields the oldest trilobites. Part of a fauna composed of "small shellies," the Tommotian is the earliest part of the Cambrian. Its strata is best represented by a sequence found along the Alden River in Siberia. This specimen is not quite that age and may be an echinoderm plate. Middle Cambrian, Big Horn Mountains, Wyoming. (Value range G).

Fig. 02-13. *Brooksella alternata*, Walcott, 1901. Considered to be fossil jellyfish by the late 19th-early 20th century paleontologist, C. D. Walcott, these "star cobbles" of the Coosa River valley are today considered by many paleontologists to be a type of trace fossil, possibly related to the above "jellyfish" *Dactyloidites*. sp. Coosa shales (Conasauga Formation), Middle Cambrian, Coosa River, Alabama. (Value range F, single specimen).

Fig. 02-14. *Camptostroma* sp. This is a problematic Cambrian fossil that has been considered by some paleontologists to have been a jellyfish, others consider it to have been an echinoderm or a Cambrian vendozoan. Lower Cambrian, Corner Brook, western Newfoundland. (Value range F).

Trace Fossils

These are trace fossils (tracks and trails) of peculiar Cambrian organisms. Some paleontologists have pointed out that the trace fossil record of the Cambrian is in some ways more diverse and varied than that of later geologic time. Diverse trace fossils appear in abundance at the same time as do fossils of animals with hard parts. Evolutionary development producing the organisms responsible for trace fossils should have left a trace fossil record showing a more gradual progression in diversity than with body fossils, since trace fossils do not require the presence of hard parts as do body fossils to leave a good fossil record. The scanty record of trace fossils prior to the Cambrian is puzzling!

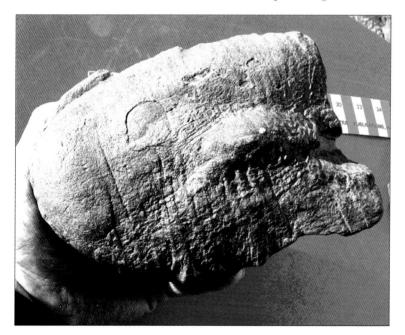

Fig. 02-15. "Big Nasty." A large trace fossil (burrow) of an equally large, burrowing animal associated with strata containing nondescript trace fossils. Such trace fossils can be ubiquitous in certain horizons of Cambrian strata, but occasionally bizarre ones like this occur with them. Its maker must have been a soft bodied, slug-like animal which pushed its way through the soft sediments of shallow Cambrian seas. In the case of this burrow, its maker was large, worm-like and possibly "nasty." Davis Formation, Upper Cambrian, Ste. Francois County, Missouri. (Value range F).

Fig. 02-16. "Circle poopers." Organic rich layers of the Upper Cambrian Davis formation of the Missouri Ozarks yield these circular sediment filled burrows. The maker burrowed into organic matter-rich mud where presumably the animal extracted organic matter upon which it fed. It then ejected the sediment which had gone through its gut, removing the organic matter as it burrowed in circles. A somewhat similar trace fossil in found in the lower Cambrian where it is known by the form genus of *Gyrolithes* sp. (Value range G).

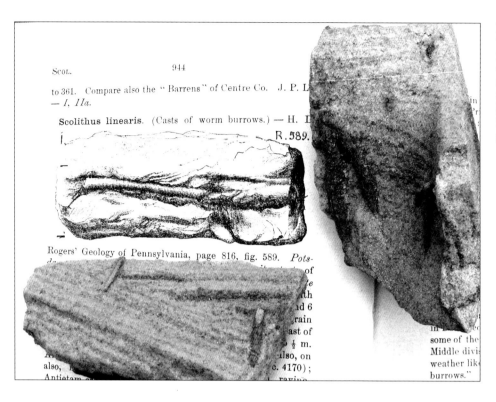

Fig. 02-17. *Scolithos* sp. These vertical "worm" burrows can occur abundantly in Cambrian sandstone. *Scolithos* (also known as *Scolithus*) is one of the earliest occurring, clearly defined trace fossils. It may have been made by a phoronid worm. Associated illustration from a late 19th century geological report.

Brachiopods

Fig. 02-18. Inarticulate brachiopod, *Obolus lamborni*. These primitive brachiopods are some of the most widespread and abundant Cambrian fossils, appearing at the very beginning of the Cambrian Period, before trilobites. Lamotte Sandstone, Ste. Francois County, Missouri.

Fig. 02-19. Articulate brachiopods, *Eoorthis remnicha*. These are the impressions of a type of brachiopod which would become very dominant throughout the rest of the Paleozoic Era. Upper Cambrian, Davis Formation, Reynolds County, Missouri.

Echinoderms

This phylum of invertebrates appear suddenly at the beginning of the Cambrian Period.

Fig. 02-20. *Golgia* sp. This is an eocrinoid, one of the more widely known and widely distributed Cambrian echinoderms. Wheeler Shale, Middle Cambrian, House Range, Utah. (Value range F).

Fig. 02-21. *Golgia* sp. Eocrinoids are early echinoderms which belong to an extinct echinoderm class, the Eocrinoidea. Unlike crinoids, eocrinoids do not have a distinct stem and their plates are more irregular. Spence Shale, Wellsville Mts., Utah. (Value range F).

24

Fig. 02-22. Echinoderm holdfasts. Hardground layers of the Upper Cambrian, Davis Formation of Missouri, locally can be covered with these small attachment structures of echinoderms. They might be the anchoring structures of crinoid-like echinoderms like that of the following photo. Davis Formation, Upper Cambrian, Ste. Francois County, Missouri. (Value range F).

Fig. 02-23. Crinoid? This undescribed echinoderm is a puzzle! It is not an eocrinoid, due to the fact that it has a distinct stem. It appears to be a crinoid, however, those Cambrian crinoids which are known lack a robust stem such as on this specimen. This echinoderm was associated with and may have been attached, to the holdfasts of the previous image. Davis Formation, Upper Cambrian, Bonneterre Missouri (Very rare).

Monoplacophorans

These fossils have been considered as monoplacophorans, a class of mollusks established in the
early 1950s, but suspected earlier from interpreta-
tion of the fossil record.

Fig. 02-24. *Shelbyoceras unguliforme,* Ulrich, Foerste and Miller.
These stubby monoplacophorans originally were considered as
early, primitive cephalopods, however, they lack the chambers
and siphuncle characteristic of a cephalopod, although they
may have been ancestral to cephalopods. These fossils, like
many other late Cambrian mollusks, are found associated
with stromatolites upon which they presumably fed. Eminence
Formation, Washington County, Missouri. (Value Range G,
single specimen).

Fig. 02-26. *Hypseloconus bessemerense.* A group of hypseloconid
monoplacophorans which may have lived clustered together,
fossilized in the position in which they fed upon stromatolite
colonies with which they are associated. Potosi Formation,
Washington Co., Missouri (Value range E).

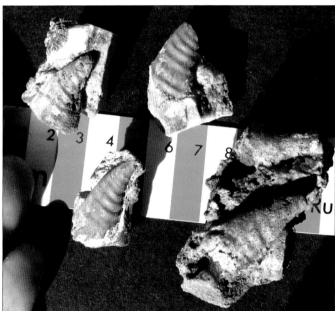

Fig. 02-27. *Gayneoconus echolsi,* Stinchcomb and Angeli.
This hypseloconid monoplacophoran has pronounced
ornamentation, a characteristic of many mollusks and
referred to as costae. The genus is named after and its
discovery associated with the music of Soviet composer Aram
Katchaturian. (Value range E for group).

Fig. 02-25. Group of Cambrian mollusks, snail-like *Scaevogyra*
and *Proplina* (middle). *Proplina* is a monoplacophoran;
Scaevogyra is considered to be an early gastropod (snail),
however, it has features that, to some paleontologists, suggest
otherwise. The white mineral at the right side of the chunk is
barite. The specimens are associated with barite occurring in
a barite pit. Potosi Formation, Washington County, Missouri.
(Value range E).

Fig. 02-28. *Hypseloconus sp.* This group of monoplacophorans are elongate hypseloconids. Monoplacophorans are a molluscan class recognized in the early 1950s from living animals dredged up from a deep sea trench (Chilean Trench) of the Pacific Ocean. They have a segmented body and a horseshoe-shaped pattern of muscle scars that reflect this segmentation. Most monoplacophorans have spoon- or cap-shaped shells; the hypseloconids, have a high, endogastrically curved shell, quite different from the spoon-shaped (patellaform) shells of modern monoplacophorans. Such high shell forms have been proposed, by some paleontologists, as representing a body plan different enough from that of modern monoplacophorans to be separated from them and included into an extinct class of mollusks, the Class Tergomya

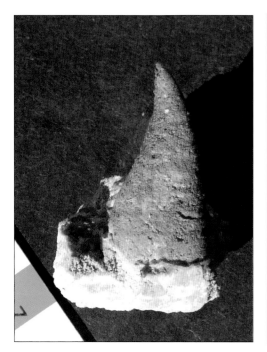

Fig. 02-29. *Hypseloconus bessemerense*, Ulrich, Foerste and Miller, 1943. This elongate monoplacophoran was originally considered to be primitive cephalopod.

Fig. 02-30. *Gayneoconus echolsi*. A single specimen of this genus of elongate monoplacophoran, which differs from *Hypseloconus* in having pronounced and distinct costae on the sides of its shell. Potosi Formation, Washington County, Mo. (Value range F).

Fig. 02-31. Seascape highlighting a group of hypseloconid monoplacophorans which cluster on the shallow sea floor of southern Missouri, 500+ million years ago. Probably such clustering would have been around stromatolites upon which they apparently fed. The Cambrian Period found widespread stromatolites in its oceans, which, as in the shallow water of the Precambrian, was a stromatolite-dominated ecosystem. Artwork by Virginia M. Stinchcomb.

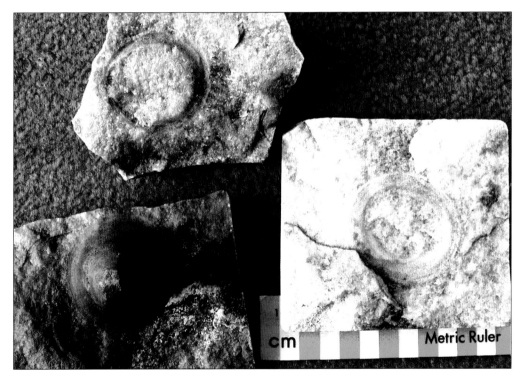

Fig. 02-32. *Palaemaecea irvingi*, Whitfield. These relatively large, cap-shaped fossils were originally described as a type of primitive gastropod (snail), which is a mollusk. They were placed into the monoplacophora in the Treatise on Invertebrate Paleontology in the 1950s. They have more recently been considered as possible cnidarans (Webers and Yochelson, 1999). Specimens such as those shown here, are preserved in quartzite and are believed by Webers and Yochelson to have been originally composed of a flexible, leather-like composition rather than having the hard, mineralized shell of a mollusk. *Palaemaecea* represents an example of a type of problematic Cambrian fossil, the Cambrian having more of such problematic forms than is the case found in younger Paleozoic strata. Wonewoc Formation, Dresbachian, Upper Cambrian, Silver Mound, Wisconsin. (Value range E).

28

Gastropods

Here are the earliest (known) undoubted gastropods (snails).

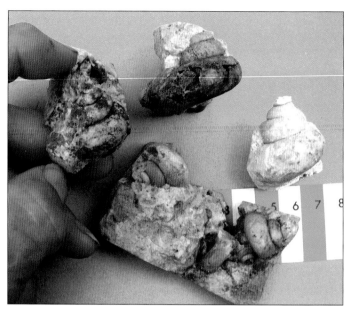

Fig. 02-33. *Dirhachopea abrupta*, Ulrich and Bridge, 1931. These represent some of the earliest undoubted gastropods (snails) that are known. Snail-like fossils such as *Scaevogyra* are known from older Cambrian strata, but all of these are suspect as to their being true gastropods. *Dirhachopea* and associated *Taeniospira* are undoubted as to their gastropod affinity. They clearly are early representatives of a large subclass of gastropods, the archeogastropoda. Eminence Formation. Late Upper Cambrian, Ste. Francois County, Missouri. (Value range F).

Fig. 02-35. *Sinuopea vera*, Ulrich and Bridge, 1931. These marine gastropods are found associated with stromatolite reefs upon which they apparently fed. *Sinuopea* and the related *Taeniospira* are among the earliest undoubted gastropods. Latest Cambrian, Van Buren Formation, Washington County, Missouri. (Value range G, single specimen).

Fig. 02-34. *Dirhachopea abrupta*, Ulrich and Bridge, 1931. This slab of impressions of the undoubted Cambrian gastropod *Dirhachopea* is relatively rare, as specimens of this early snail are usually found separate rather than clustered together as they arc here. These gastropods, like most mollusks of the Cambrian and the Lower Ordovician, lived associated with stromatolites. Such a stromatolite-dominated ecosystem in the early Paleozoic was a relic from the Precambrian where stromatolites were often the only obvious evidence of life. (Value range F).

Fig. 02-36. *Taeniospira* sp. This somewhat high spired gastropod is the genus *Taeniospira*. Diversity of species in Cambrian gastropods is not very great. The various species described for *Taeniospira* and *Dirhachopea* is questioned by some paleontologists, including Ellis Yochelson, an authority on early snails (personal communication). Eminence Formation, Womack, Missouri.

These fossils look like (and possibly were) gastropods, however they have some peculiar, un-gastropod-like characteristics about them such as a planispiral shell.

Fig. 02-37. *Macluritella walcotti*, Howell, 1946. These planispiral molluscan fossils were originally described as a type of *Hyolithes*, but later investigation recognized them as a type of very open-coiled gastropod. Eminence Formation, Shannon Co., Missouri. (Value range F for single specimen).

Fig. 02-38. *Strepsodiscus* sp. This planispiral gastropod or gastropod-like mollusk is a typical Cambrian fossil. Like many Cambrian fossils, they occur associated with stromatolites, in this case with digitate stromatolites of the early Upper Cambrian. Some specimens of this gastropod have been found in the lead mines of southeastern Missouri where the molluscan shell has been replaced or preserved with galena. Bonneterre Dolomite, Irondale, Missouri. (Value range G, single specimen).

Fig. 02-39. These snail-like fossils come in left- and right-handed varieties. They have been questioned for this reason as to their gastropod affinities. Most Cambrian snail-like fossil shells such as *Plagiella* and *Scaerogyra* are suspect as to their being true gastropods. Franconia Formation, Taylor's Falls, Minnesota. (Value range G).

These Onychochilid "gastropods" are puzzling. Nothing is known about their soft parts and there is no definitive reason why they could not be classified as gastropods; they just don't match-up with undoubted, early gastropods. They are odd!

Everything here is a proper snail (gastropod).

Fig. 02-40. *Scaevogyra swezeyi*, Whitfield. This group of gastropod-like, left-handed coiling mollusks is puzzling. They occur with a variety of elongate, cone-shaped monoplacophorans associated with digitate stromatolites. Potosi Formation, ashington County, Missouri. (Value range G. single specimen).

Fig. 02-42. *"Hamatospira"* sp.These small, loosely coiled gastropods were suggested to have lived in cavities within stromatolites by Ellis Yochelson (personal communication) and to have been super sluggish snails. Eminence Formation, Potosi Missouri. The genus is not a valid, officially described and recognized genus.

Fig. 02-41. *Scaevogyra S. "saxa."* These small, left-handed gastropod-like fossils belong to the genus *Scaevogyra*, which is a representative of the peculiar gastropod(?) family Onychochilidae. This species *Scaevogyra saxa*, was described in a thesis and, hence, it is technically not a valid paleontological species. Eminence Formation, Potosi, Missouri.

Fig. 02-43. Seascape of Late Cambrian, shallow water with early gastropods (*Taeniospira*), trilobites (*Stenopilus*), and *Calvinella*. All of these organisms shown living between stromatolite reefs. Artwork by V. M. Stinchcomb.

Hyoliths

Hyoliths are common and problematic Cambrian fossils.

Fig. 02-44. *Hyolithes* sp. This chert concretion contains impressions of clustered hyolith specimens, a problematic and common Cambrian organism. Some paleontologists regard *Hyolithes* as representative of an extinct class of mollusks, while others consider them as members of an extinct phylum and not a mollusk. Middle Cambrian, Conasauga Formation, Coosa River, Georgia. (Value range F).

Trilobites

Olenellus is one of the earliest trilobites, it appears in the Lower Cambrian as part of the North American faunal province.

Fig. 02-45. *Olenellus gilberti*. The Olenellids are the earliest trilobites normally seen in collections. They belong to a primitive order of trilobites, the Redlichida, which includes the genera *Olenellus*, *Paedeumias*, and *Nevadia*. Pioche Shale, C Member, Pioche, Nevada. (Value range F, single specimen).

Moroccan trilobites come from the Avalonian, Baltic or North Atlantic faunal province of the Cambrian Period.

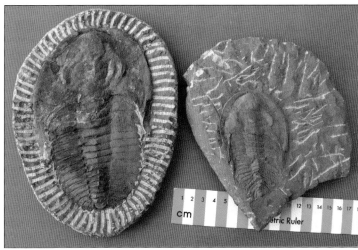

Fig. 02-46. *Cambriopallis (Andusilina) telesto*, with radial chisel marks (left), small specimen with little or no preparation (right), a type of Moroccan trilobite often preferred by some collectors. Avalonian Province trilobites from the Cambrian of Morocco, as well as elsewhere, can be quite large. This early trilobite has been collected in large numbers from brown siltstone beds in the Atlas Mountains where relatively large quarries for the collection of fossils have been opened in the trilobite bearing layers. Complete, large Cambrian trilobites are usually hard to get, these, widely distributed trilobites, are a notable exception. (Value range E, single specimen).

Fig. 02-47. *Cambriopallis (Andusilina) telesto*. Three exceptionally large specimens of this widely distributed trilobite from youngest Lower Cambrian strata of the Atlas Mountains of Morocco. Rumors have circulated that these large Moroccan trilobites are man-made replicas (casts) and are not real specimens. The author has not found this to be the case, however, some specimens have been extensively restored and sometimes over-prepared. At the Atlas Mountain localities where these trilobites are quarried, they occur in such abundance that they can be sold at low prices. Totally unprepared specimens of these trilobites are often more desirable than are the ones that, to a novice, are "more attractive" because of extensive preparation and "restoration." (Value range E).

Fig. 02-48. *Cambriopallis (Andusilina) telesto.* An "overprepared" specimen of this interesting trilobite from Morocco. Some of the specimens coming from Morocco are prepared in such a way that second-rate specimens are reconstructed to be more impressive and hence more saleable. Such specimens are not the same as replicas, even though part of the trilobite has been reconstructed. Specimens which are highly prepared show no overlapping of the thoraxial segment impressions; lightly or unprepared specimens do. (All of the Moroccan Cambrian trilobites are impressions.) Such replicas or "fakes" can be identified as such by the use of a hand lens or binocular microscope. If it looks too perfect it is probably an "over prepared" specimen but not necessarily a "fake." (Value range F, as overprepared "replica").

Fig. 02-49. *Cambriopallis* sp. Close up of carapace impression of the specimen in 02-47. In this original, the impressions of the animals exoskeleton or carapace can be seen. Note that in this natural mold, no original material is preserved, the place where the exoskeleton was originally present has left a clear impression in the matrix. Replicas and related overprepared specimens, do not show such detail.

Fig. 02-50. Somewhat enlarged from the previous view.

Fig. 02-51. Another view of the impression of a *Cambriopallas* carapace. Note that thoraxial segments on unaltered specimens are impressions as no original material is preserved in the Moroccan Cambrian siltstones. A natural, yellow ocher coats the mold of the trilobite.

Common trilobites of the North American faunal province of the Cambrian Period.

Fig. 02-53. *Elrathia kingi*, Meek. The fourth specimen on top right has a "bite mark" on it. Such bite marks are found on these well known trilobites from the House Range of Utah and are considered as evidence of predation. Such predation may have been by a large arthropod found in Cambrian trilobite zones known as *Anomalocaris*. Wheeler Shale, House Range, Utah.

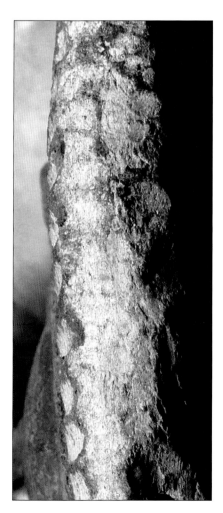

Fig. 02-52. *Cambriopallas telesto*. Close up of the edge of *Cambriopallas* "replica" or overprepared specimen. Note the granular material making up the left side of this specimen; this is not the natural rock. Part of the specimen is a resin cast of part of a trilobite which has been cemented to a slab of siltstone which composes the right part of this specimen.

Fig. 02-54. *Elrathia kingi*, Meek. Close-up of the same specimen as above showing the healed bite mark on the specimen in the previous image. Wheeler Shale, House Range, Utah. (Value range F).

34

Fig. 02-55. *Elrathia* sp. Cambrian trilobites similar to those of western Utah occur in a "down dropped fault block" at the eastern end of Pend Oreille Lake in northern Idaho.

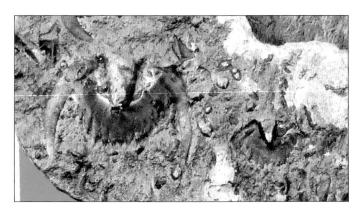

Fig. 02-57. *Drepanura* sp. A distinctive Upper Cambrian trilobite from one of the Cambrian faunal provinces of China. Changxia, Shandond, China.

Trilobites from Cambrian faunal provinces of Asia.

Trilobites from the Baltic, Avalonian, or North Atlantic Province.

Fig. 02-56. Unidentified trilobite partials, Alaska. Many of the Paleozoic fossils from Alaska do not match up with similar age specimens from other parts of North America. This is because a large portion of Alaska is made up of what is known as autochronous terrain. Such terrain represents crustal rocks which were formed somewhere else on the earth and later transported by sea floor spreading to another part of the planet and then added or "welded" to the place where they occur today. These Upper Cambrian trilobites are from such autochronous terrain and represent a Cambrian faunal province best represented in Kazakhstan and known as the Kazakhstanian faunal province. Most Cambrian trilobites found in North America belong to the North American or Laurentian faunal Province, however on the eastern part of North America autochronous terrain of the Avalonian or North Atlantic province occurs, often with large trilobites. Middle Cambrian, Snowden Mountain, near Livingood, Alaska.

Fig. 02-58. *Acadioparadoxides briareus.* This large trilobite is a paradoxid, a family of large trilobites of which *Paradoxides* is the best known genus. These trilobites belong to the Avalonian or North Atlantic faunal province of the Cambrian, which is quite different in its faunal composition than similarly aged Cambrian fossils of the North American or Laurentian faunal province. This faunal province (Avalonian) was named after eastern Newfoundland's Avalon Peninsula, where well preserved trilobites occur in its slaty rocks. (Value range E).

Fig. 02-59. *Acadioparadoxides briareus.* These are some of the largest trilobites of the Cambrian; they are trilobite giants! The zone carrying these trilobites has been quarried to produce specimens in large numbers from Middle Cambrian ferruginous siltstone beds of the Atlas Mountains of Morocco and as of this writing are readily available. Such windows of opportunity don't always remain open, so a serious trilobite collector should try to obtain one of these sweeties. (Value range E).

Trilobites from Upper Cambrian strata are usually not complete, The Upper Cambrian is, however, the time of the greatest diversity in trilobites.

Fig. 02-60. *Lonchopygella mansuyi*, Kobayashi. This trilobite is representative of the Chinese Yunnan faunal province of the Cambrian period. It represents one of some seven distinct faunal provinces of the Cambrian Period. Upper Cambrian, Yunnan Province, China.

Fig. 02-61. A seascape with a group of the early late Cambrian (Dresbachian) trilobite *Tricrepicephalus* swimming in a shallow, Cambrian seaway. Artwork by Virginia M. Stinchcomb.

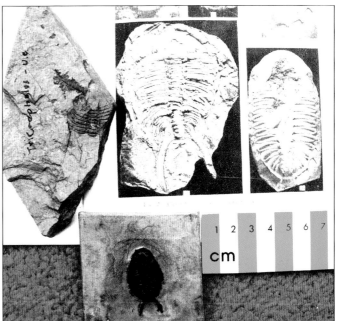

Fig. 02-62. Specimens of the trilobite *Tricrepicephalus* from now submerged Cedar Bluff, Alabama. Slaty, crumbly shale of the late Cambrian (Dresbachian) stage from Cedar Bluff was the source of nice Dresbachian trilobites. The black and white photos of Cedar Bluff specimens are from "The Geology of Alabama" by Charles Butts, an early 20th century geologist/ paleontologist who worked in the southern Appalachians.

Fig. 02-63. *Tricrepicephalus texanus*. A group of cranidia from this trilobite "marker" of the earliest Late Cambrian. Bonneterre Formation, Ste. Francois County, Missouri. (Value range E for group).

Fig. 02-64. *Prosaukia* sp. These late Cambrian saukid trilobites are from the Llano Uplift in the Texas hill country near Fredericksburg, Texas. Most of Texas is underlain by Mesozoic rocks, however in the Llano Uplift, tectonic forces have brought up much older rocks of Precambrian and Cambrian age that otherwise, in Texas, are deeply buried beneath younger strata. Wilberns Formation, Upper Cambrian near Fredericksburg, Texas.

Fig. 02-65. *Pseudolisania breviloba*, Walcott. The Appalachian mountains expose a great deal of Cambrian and older strata. These trilobite pygidia come from slaty shales of the Upper Cambrian, Nolichucky Formation of eastern Tennessee. (Value range G).

Fig. 02-66. *Coosia* sp. This group of trilobite cranidia and pygidia come from Cambrian limestone beds of the Black Hills near Lead, South Dakota. Like the Llano Uplift of Texas, mentioned above, the Black Hills bring to the surface strata of Precambrian and Cambrian age that, in other parts of South Dakota, are deeply buried by younger layers, most of which are of Mesozoic age. Cambrian strata overlie Precambrian rocks of the Black Hills with a major unconformity (hiatus) between the two series. Prior to the Cambrian, the surface of this part of the earth was subjected to millions of years of weathering. This weathering concentrated gold that was originally in the Precambrian black slate beds near what would become Lead (pronounced Leed), South Dakota. This gold-rich product-of-weathering became part of the Cambrian gold bearing conglomerate beds which formed the basis of the Black Hills gold rush in the late 19th century. The gold bearing, Precambrian black slates became the site of the Homestake Gold Mine near Lead. The beds yielding these trilobites lie above the (mined out) gold bearing conglomerates of Cambrian age.

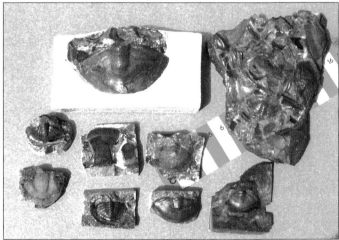

Fig. 02-67. *Coosia* sp. Another group of trilobite partials from the Cambrian of the Black Hills, South Dakota. This trilobite genus is named for the Coosa River of Georgia and Alabama from where this trilobite was first collected

Fig. 02-68. *Ataktaspis* sp. The Upper Cambrian was a time of maximum diversification of trilobites, however many of them are small and most are not found complete, but are found in thin limestone beds as partials such as these triangular shaped pygidia (tails). The diversity of Upper Cambrian trilobites can be overwhelming, but they provide excellent biostratigraphic zonation where they occur. Bonneterre Formation, Lower Upper Cambrian (Dresbachian), Ste. Francois County, Missouri. (Value range G, single specimen, E for group).

Fig. 02-69. *Stenoplius* sp. A trilobite characteristic of the late Cambrian of the North American Province found associated with areas of extensive stromatolite reefs. *Stenoplius* has a large, hemispherical cephalon (specimens shown to the right of complete ones) and the complete trilobite resembles a large pill bug. Eminence Formation, Potosi, Missouri, (Value range E, for complete specimens, F for partials).

Fig. 02-72. The trilobite *Calvinella ozarkensis* on a clear, shallow sea floor of the late Cambrian. Artwork by Virginia M. Stinchcomb.

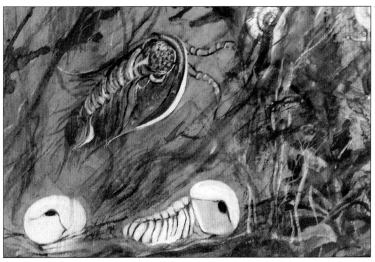

Fig. 02-7. Two *Stenopilus* specimens living and feeding(?) upon nearby stromatolites. Artwork by Virginia M. Stinchcomb.

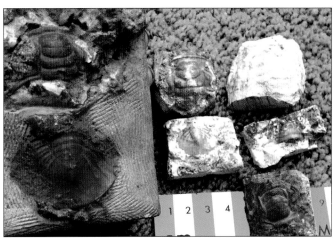

Fig. 02-71. *Calvinella ozarkensis*, Ulrich and Bridge, 1931. *Calvinella* is a trilobite that belongs to the Saukids, a family of late Cambrian and early Ordovician (Ozarkian) trilobites, some of which can get quite large. Here is a group of *Calvinella* cranidia and specimens of the characteristic, fan shaped pygidia. Eminence Formation, Washington County, Missouri. (Value range F for single cephalon or pygidia.).

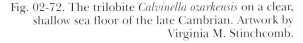

Trilobitomorph Trace Fossils.

These trace fossils are believed to have been made by trilobitomorphs which were soft bodied, trilobite-like arthropods.

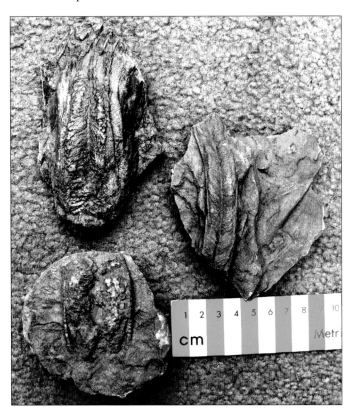

Fig. 02-73. *Cruzania* sp. These are the sediment fillings of what are known as trilobite resting pits. Trilobites would dig a two-sided pit in appropriate mud on the sea floor. Such a "pit" could later be filled with sand thus making a sand mold of the pit. Such trace fossils are given the form genus *Cruzania* and *Rusophycus* depending upon their length. Specimen on the right is *Rusophycus* sp. Gros Ventre Formation, Middle Cambrian, Big Horn Mts., Wyoming.

Fig. 02-74. *Cruzania* sp. These numerous small trilobite resting pits were made by a group of small trilobites doing the same thing as was done by their more mature trilobite associates who made the trace fossils of the previous image. This obviously was an instinctive rather than a learned behavior, as the "hatchlings" which made these small pits could not have learned it; an example of fossilized innate behavior over 500 million years old. Gros Ventre Formation, Middle Cambrian, Big Horn Mts., Wyoming.

Graptolites.

Fig. 02-75. Denderitic graptolites. *Denderograptus* sp. Graptolites are extinct representatives of the phylum Chordata, subphylum hemichordate. These bottom-dwelling forms were typical of the Cambrian Period. In later periods of the Paleozoic pelagic (floating forms) became the norm. Upper Cambrian, Franconia Formation, Afton, Minnesota. (Value range F, single specimen).

Fig. 02-76. Black Hills outcrop of Cambrian strata, Black Hills, South Dakota. Lower brown beds are thick sandstone layers with overlying Cambrian dolomite beds. The meadow in the foreground is underlain by Precambrian slate.

Bibliography

Gould, Steven J. "Is the Cambrian Explosion a Sigmoid Fraud?" in *Ever Since Darwin. Reflections in Natural History.* W.W. Norton and Company, New York, London, 1977.

_____. "The Pentagon of Life" in *Ever Since Darwin. Reflections in Natural History.* W.W. Norton Company, New York, London, 1977.

_____. "An Unsung Single-Celled Hero" in *Ever Since Darwin. Reflections in Natural History.* W.W. Norton Co., New York, London, 1977.

_____. *Wonderful Life. The Burgess Shale and the Nature of History.* W.W. Norton and Co., New York-London, 1969, .

Morris, Simon Conway. *The Crucible of Creation. The Burgess shale and the rise of Animals.* Oxford University Press, 1998.

Robison, Richard A. "Middle Cambrian Biodiversity Examples from Utah Lagerstatten," in *The Early Evolution of Metazoans and the Significance of Problematic Taxa.* Simonetta and Morris, Simon C. Cambridge University Press, New York-London, 1989.

Yochelson, Ellis and B. Stinchcomb. "Recognition of Macluritella (Gastropoda) from the Upper Cambrian of Missouri and Nevada," *Journal of Paleontology,* Vol. 61, p. 56-61, 1987.

Chapter Three

Lower Ordovician (Ozarkian and Canadian)

A Different Part of the Ordovician

If there is anything in this work that is going to upset purists it might be this separation of the Lower Ordovician period from the rest of the period. This is being done because the Lower Ordovician has distinct fossils and sedimentational features, a phenomena noticed by many geologists and paleontologists of the past.

Probably the best known attempt at isolating the Lower Ordovician and separating it into its own geologic period (or periods) was that of stratigrapher/paleontologist E. O. Ulrich in the 1910s through the 1930s. Ulrich established a geologic period that he called the Ozarkian, which represented the time (and its representative strata) from the end of the Cambrian (later to include part of the Upper Cambrian) well into the Lower Ordovician. He also proposed that the latter (younger) half of the Lower Ordovician would become another geologic period which he called the Canadian. Thus he proposed the addition of two geologic periods carved out of the latest part of the Cambrian and the oldest part of the Ordovician, the Ozarkian and the Canadian. Despite Ulrich's considerable ego, or perhaps because of it, the change did not win acceptance in the highly politicized world of the North American geologic community.

However, the fact remains that, worldwide, the earlier part of the Ordovician Period, is quite different in both its fossils and strata from the later part of the period. One of the reasons for this is that during the Lower Ordovician the seas which covered the craton were still full of stromatolites, a vestigial ecosystem that harkened all the way back to the Precambrian. This abundance of stromatolites formed the basis of an ecosystem that would disappear after the end of the Lower Ordovician (the late Canadian).

After the Lower Ordovician, the biomass of lime-secreting invertebrates like corals, brachiopods, and bryozoans increased to the point that they may have taken over much of the job of calcium carbonate precipitation previously done by stromatolites. Separation of the Lower Ordovician from the rest of the period thus makes sense from a paleoecological context.

The Lower Ordovician has recently been named the Ibexian after Ibex, Utah, where a complete section of fossiliferous limestone of Lower Ordovician age occurs. It was formed in what was then the edge of the North American craton (or continent).

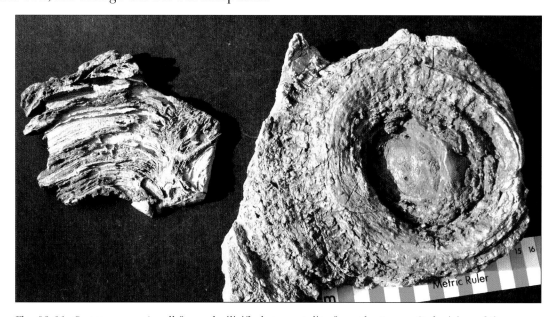

Fig. 03-01. *Cryptozoon* sp. A well formed, silicified stromatolite from the Lower Ordovician of the Ozarks. Stromatolites are structures produced by a community of monerans; they represent an ancient biogenic community that goes back to the early history of the earth. Cotter Formation, Taney County, Missouri (Value range F)

Stromatolites

Stromatolites dominate much of the strata formed in the shallow waters of the craton during the Lower Ordovician. Sponges and other lower invertebrates can be found as fossils where they were associated with this stromatolite dominated ecosystem.

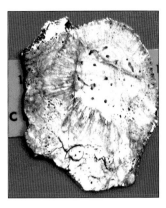

Fig. 03-02. *Archeoscyphia annulata,* Cullison, 1944. This sponge lived among stromatolites in what would become the Ozark region of Missouri and Arkansas, during the middle part of the Lower Ordovician. These sponges, like most others, were delicate; they are preserved associated with beds of chert and chert nodules. They generally are not conspicuous and can be subtile fossils. Douglas County, Missouri (Value range G).

Fig. 03-03. *Nicholsonella* sp."Nothing fossils." These silica (quartz) replaced fungi-looking fossils represent one of a number of problematic fossils found in early Paleozoic strata. Once a paleontologic enigma, these fungus-like growths which are associated with stromatolites, are now believed to be an early type of bryozoan. Smithville Formation. Stoddard County, Mo. (Value range F).

Fig. 03-04. "Worm tracks." Cambrian and Lower Ordovician (Ozarkian) slabby sandstone layers can have peculiar trace fossils on their bedding surfaces. Such "worm tracks" are a relatively common type of trace fossvil. Roubidoux Formation, Phelps Co., Mo. (Value range G)

Mollusks

Mollusks were particularly successful in the stromatolite-dominated ecosystem of the Lower Ordovician. One of the successful groups were the monoplacophorans, a molluscan class represented sparingly in today's oceans.

Fig. 03-05. Multiplated mollusks. Variously shaped valves or plates of molluscan origin are found in Ozarkian strata and can be locally abundant. Such fossils are considered as early chitons by some paleontologists while others (including the author) consider them to be the valves or plates of an extinct molluscan body plan, that is an extinct molluscan class. (Value range E for group).

Fig. 03-06. *Gasconadeoconus ponderosa*, Stinchcomb and Angeli. Many Ozarkian fossils (and the animals responsible for them) were very similar to those of the Cambrian Period. These cone-shaped monoplacophorans are similar but larger than similar Cambrian forms. Large size is a characteristic of many Ozarkian mollusks where they were associated with stromatolite reefs. Gasconade Formation, Crawford Co., Miss

Fig. 03-07. Another group of these large, cone-shaped monoplacophorans. Small, similar shaped monoplacs found in Cambrian rocks belong to the genus *Kirengella*. These large "monoplacs" also have a distinctive shell ornamentation, which is not seen on these internal molds or steinkerns. Gasconade Formation, Crawford Co., Missouri.

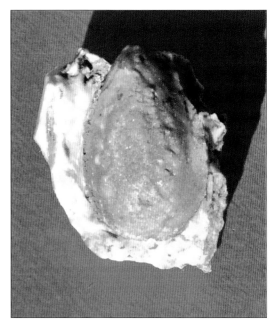

Fig. 03-08. *Bipulvina croftsae*, Yochelson. Spoon-shaped monoplacophorans are the most typical shell form of this class of mollusks. Monoplacophorans are well represented in the late Cambrian and the early Ozarkian when they were at the zenith of their diversity. The author of the genus was Ellis Yochelson (1928-2006), a late 20th-early 21st century researcher on early mollusks and on early problematic fossils. The horseshoe-shaped pattern of multiple muscle scars, seen here, is what distinguishes fossil monoplacophorans from other mollusks. This pattern of muscle scars, reflects segmentation of the mollusks soft parts, a feature not found in other molluscan classes. Gasconade Formation, *Bipulvina* zone, early most Ozarkian. (Value range F).

Fig. 03-09. *Titanoplina* sp. These proplinid (spoon-shaped) monoplacophorans can be over six inches in length, a truly gargantuan size for monoplacophorans, which usually are relatively small. Gasconade Formation, Pulaski County, Mo. (Value range F, single specimen).

Fig. 03-11. *Proplina meramecensis*. A group of medium-sized monoplacophorans with very faint muscle scars associated with relatively large gastropods, (bottom slab). Gasconade Formation, Washington Co., Mo.

Fig. 03-10. *Proplina sibeliusi*. These proplinid monoplacophorans had a relatively thin shell so that muscle scars are not evident. Musculature is usually associated with thick molluscan shells. The species is named after Jean Sibelius (1865-1961), Finnish symphonic composer. Gasconade Formation, Crawford Co., Mo.

Fig. 03-12. *Proplina meramecensis*. Another group of cap- or spoon-shaped monoplacs from lowermost Ordovician strata of the Ozarks. (Value range F, single specimen).

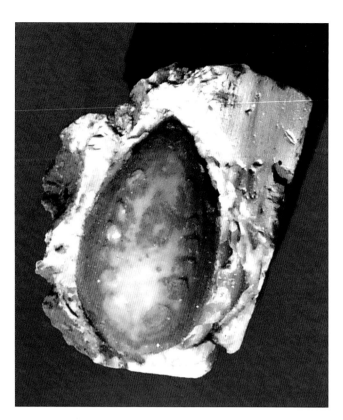

Fig. 03-13. *Proplina cornutiformis*. This relatively common and widespread spoon-shaped monoplacophoran can have distinct and well formed muscle scars on some (but not all) specimens. Pronounced muscle scars like those shown here are unknown on elongate forms such as *Hypseloconus* and are rare on the cone-shaped forms such as *Kirengella*. Gasconade Formation, Crawford Co., Mo. (value range F).

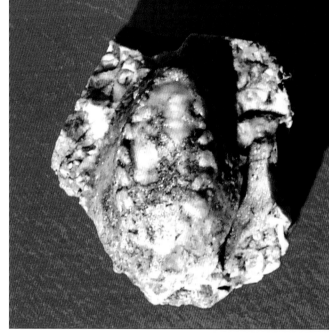

Fig. 03-14. *Proplina cornutiformis*. Muscle scars (or musculature) is quite pronounced on this specimen from a zone near the base of Ozarkian strata. They are almost Cambrian in age and may well be Cambrian, depending upon where in the rock strata in which they were found the Cambrian-Ordovician boundary is drawn. All of these morphologic features are preserved on chert steinkerns which is capable of preserving features on fossils with great fidelity and detail.

Fig. 03-15. *Proplina cornutiformis*. Another group of these fossil mollusks, fossils of the class monoplacophora. Seldom in a "popular," non-technical work like this one are so many representatives of this molluscan class shown. I shall break a record in showing a number of examples of these interesting fossil mollusks.

Gastropods

Gastropods or snails were one of the other successful molluscan classes of the shallow, stromatolite-dominated waters of the Lower Ordovician.

Fig. 03-16. *Helicotoma* sp. and *Eumophalopsis* sp. (small). These gastropods occur in chert layers which preserve fossil mollusks with considerable fidelity. They are found in pockets throughout the Ozarks, but similar fossils are found in the southern Appalachians as well as elsewhere where strata of this age occur. These mollusks fed on and lived associated with stromatolites in the Lower part of the Ordovician period. This was a time which represented the last of this ancient stromatolite-dominated ecosystem. Washington County, Missouri (Value range F).

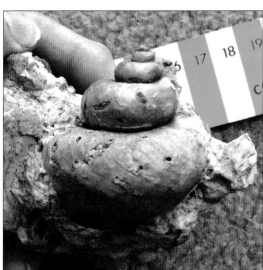

Fig. 03-17. *Sinuopea regalis*, Ulrich and Bridge, 1931. This is a large, early Ozarkian gastropod. Undoubted gastropods first appear in the Late Cambrian. At the beginning of the Ordovician they become larger although still similar in form to their Cambrian ancestors. Gasconade Formation, Lower Ozarkian, Crawford Co., Missouri. (Value range F).

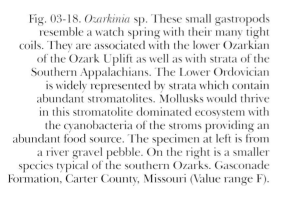

Fig. 03-18. *Ozarkinia* sp. These small gastropods resemble a watch spring with their many tight coils. They are associated with the lower Ozarkian of the Ozark Uplift as well as with strata of the Southern Appalachians. The Lower Ordovician is widely represented by strata which contain abundant stromatolites. Mollusks would thrive in this stromatolite dominated ecosystem with the cyanobacteria of the stroms providing an abundant food source. The specimen at left is from a river gravel pebble. On the right is a smaller species typical of the southern Ozarks. Gasconade Formation, Carter County, Missouri (Value range F).

46

Fig. 03-19. *Lecanospira compacta*, Ulrich and Bridge. These gastropods, with their sinistral coiling (left-handed coiling), occur frequently in groups like those shown here. *Lecanospira* can be locally abundant in the southern Ozarks of Missouri and slabs like this have been widely distributed. In the 1950s, local rock shops in the Ozarks would sometimes sell slabs of them as tourist souvenirs. One of the sources of *Lecanospira* slabs were rock shops along Route 66 between Rolla and Springfield, Missouri. You could get your "Kicks on Route 66" with these fossils as well as with other attractions. Roubidoux Formation, Berryman, Missouri. (Value range E).

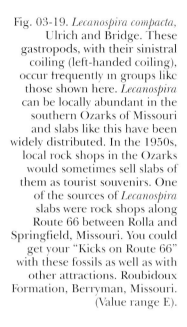

Fig. 03-20. *Lecanospira compacta*. The shell of *Lecanospira* is flat on the top (anterior) and concave on the bottom (posterior). These are relatively large specimens of this widespread and distinctive Lower Ordovician gastropod. Roubidoux Formation, Oregon County, Missouri.

Fig. 03-21. *Lecanospira compacta* These groups of *Lecanospira* are quite nice and certainly pleasing to the eye!

Fig. 03-22. *Leseurilla isabellaensis*, Cullison. This loosely coiled genus of gastropod is characteristic of the middle part of the Lower Ordovician. Many Ozarkian (or Lower Ordovician) gastropods and cephalopods are loosely coiled like this form. Cotter Formation, Taney County, Missouri.

Fig. 03-23. *Orospira elegantula*, Cullison. These early gastropods are unusual in being so highly ornamented. Most Cambrian and Ordovician gastropods lack such ornamentation. These are impressions of the gastropod shells in chert; original shell material is usually not preserved on Ozarkian and Canadian fossils.

Fig. 03-25. *Murchisonia (Turritoma)* sp. Close-up of slab on the left seen in the previous image. Smithville Formation, Southeast Missouri. (Value range F).

Fig. 03-26. *Murchisonia (Turritoma) acrea*. From their first appearance in the Late Canadian (Lower Ordovician), high spired gastropods will become dominant mollusks in the sea from the Middle Ordovician until the present. Smithville Formation, Bollinger Co., Missouri.

Fig. 03-24. *Murchisonia (Turritoma)* sp. High spired gastropods like this appear for the first time in the Late Canadian. These nice specimens are silicified, that is, their original shell material has been replaced with silica (chert). Smithville Formation, Bollinger County, Mo.

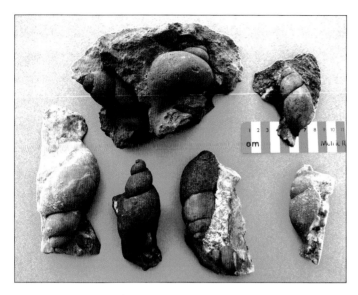

Fig. 03-27. *Subulites sp.* A delectable group of different species of these large, high spired gastropods from the late Lower Ordovician (Late Canadian) strata associated with a reef of large stromatolites. Smithville Formation, Bollinger Co., Missouri.

Fig. 03-28. *Ceratopea unguis.* A thick and robust cap fit onto the anterior of the animal which occupied a snail shell; such a cap is called an operculum. These can be plentiful fossils in late Lower Ordovician strata, specifically that strata known as Canadian in age. Species of *Ceratopea* can be excellent biostratigraphic markers (fossils used as time indexes in strata) in Canadian strata of North America and in that part of Europe which contains North American plate geology (Greenland and Scotland). Smithville Formation, Bollinger County, Mo.

Fig. 03-29. *Clathrospira subconica*, Hall. In the Late Canadian, gastropods and cephalopods diversify. Forms like *Clathospira*, offer a glimpse of what will be in store in the "snail world" in the geologic future, which in this case would be the Middle and Upper parts of the Ordovician Period. Smithville Formation, Bollinger Co., Mo. (Value range G, single specimen).

Cephalopods

These are the earliest cephalopods. They come from the earliest part of the Lower Ordovician or the Ozarkian. Cephalopods of later geologic time would become the highly evolved mollusks that today includes the octopus, squid, and pearly nautilus.

Fig. 03-30. *Dakeoceras subcurvatum*. These are some of the earliest cephalopods. They occur near the Cambrian-Ordovician boundary and depending upon where the boundary is drawn, might even be considered Cambrian in age. They are associated with Cambrian-type fossils such as kirengellid monoplacophorans and plated mollusks. Van Buren Formation, Potosi, Mo. (Value range G, single specimen).

Fig. 03-32. *Clarkoceras* sp. A early cephalopod which still retains the silicified shell so that the shell's closely spaced chambers are not visible. These early cephalopods, known as ellesmeroids, resemble elongate monoplacophorans from which they probably evolved. Gasconade Formation, Waynesville, Missouri. (Value range F).

Fig. 03-31. *Dakeoceras subcurvatum*. A natural group of complete, early cephalopods found as a stromatolite "core." Shown here are living chambers plus the thin, tightly packed chambers that give these early cephalopods a segmented look. Van Buren Formation, Potosi, Mo. (Value range F).

Fig. 03-33. Group of brevicone, ellesmeroid cephalopods. The Lower Ozarkian of the southern Appalachians, the Ozarks and the upper Mississippi Valley has one of the largest and most extensive early cephalopod faunas in the world. These specimens are from the Lower Ozarkian of the Ozarks. (Value range F, single specimen).

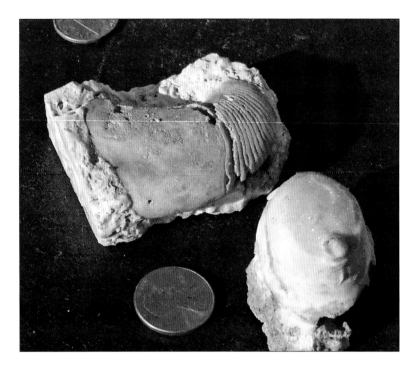

Fig. 03-34. *Oneotoceras percurvatum*. An early cephalopod with a "crimped" living chamber. This is an ellesmeroid cephalopod, a member of an early and primitive cephalopod family. The specimen at 5:00 shows the shell's siphuncle; the top specimen shows the closely spaced chamber walls characteristic of ellesmeroids. Gasconade Formation, Crawford Co., Mo.

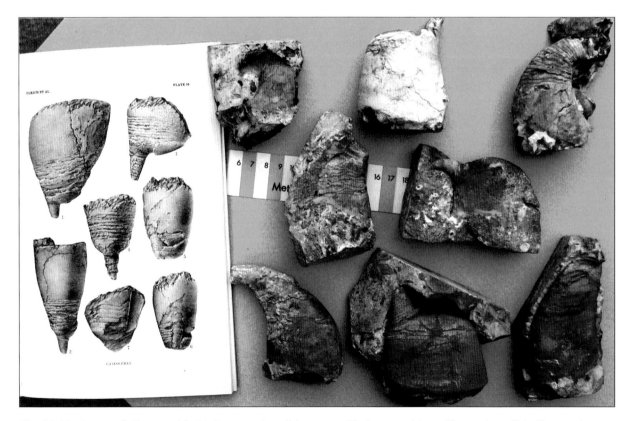

Fig. 03-35. Group of ellesmeroids. Various species of the genus *Clarkoceras* with an illustration of similar specimens from "Ozarkian and Canadian Cephalopods," a 1940s publication about them where they were first thoroughly covered in the scientific literature. All of the earliest cephalopods have these slightly curved shells with closely spaced chambers in the phragmocone (that part of the cephalopod shell not occupied by the living chamber). (Value range F, single specimen).

Ozarkian straight, nautaloid cephalopods.

Fig. 03-36. *Pachendoceras* huzzahense, Ulrich and Foerste. One of the earliest of the straight cephalopods, a group which would range through the entire Paleozoic before going extinct at the end of the Permian. Here can be seen vestiges of their beginning. Gasconade Formation, Crawford Co., Missouri (Value range F, single specimen).

Fig. 03-37. *Pachendoceras huzzahense*. Plate at right illustrates specimens from the same general area as are the actual specimens. The plate came from a work on Ozarkian (and Canadian) fossils.

Fig. 03-38. Another group of early, straight cephalopods of the type found at the very beginning of the Lower Ordovician. They are accompanied by the title page of *Ozarkian and Canadian Cephalopods*, a 1942 work on these early mollusks.

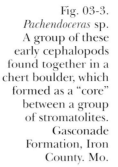

Fig. 03-3. *Pachendoceras* sp. A group of these early cephalopods found together in a chert boulder, which formed as a "core" between a group of stromatolites. Gasconade Formation, Iron County. Mo.

Fig. 03-40. A group of *Pachendoceras* specimens set in plaster and oriented in the manner and position in which they were found associated between two large stromatolites. Pulaski Co., Mo.

Fig. 03-41. Eremoceras sp. Two specimens of this ellesmeroid cephalopod showing the living chamber and the phragmacone made up of many closely spaced chambers. Roubidoux Formation, Wilderness, Missouri, (Value range F).

Fig. 03-42. *Mcqueenoceras jeffersonense*. These are the siphuncles of an early straight cephalopod. The siphuncles can be concentrated upon weathering of the surrounding dolomite with only the siphuncle being preserved. They resemble the butts of filter king cigarettes with which they can be confused and they sometimes can be collected from gullied out road ditches where cigarette butts can also lurk. Jefferson City Formation, Phelps Co., Missouri. (Value range H, single specimen).

Examples of the earliest coiled nautaloid cephalopods. They first appear in the late Ozarkian, but most early coiled nautaloid cephalopods are Canadian in age.

Fig. 03-43. *Aphetoceras* sp. One of the earliest coiled cephalopods. The first cephalopod shells are slightly curved (brevicones), followed by straight shell forms. These are followed by coiled forms, the earliest of which are disjunct, that is the living chamber separates from the rest of the shell near the outer part of the shell or, as the case here, the whorls of the entire shell are never really touching. Jefferson City, Rolla, Missouri. (Value range F).

Fig. 03-44. An assemblage of mollusks and a sponge forming a "core" that occupied the center of a group of stromatolites. At the bottom left is a monoplacophoran of the genus *Proplina*. A coiled cephalopod is in the center and a sponge to the right of that. Such an assemblage may represent the position of these animals when they were living. Cephalopods and other mollusks were associated with or sheltered by large stromatolites with which this "core" and other similar Ozark molluscan occurrences are associated. Quarry Ledge, Jefferson City (Rich Fountain) Formation, Gainesville, Mo. (Value range E).

Fig. 03-45. A coiled nautaloid cephalopod in a chert concretion which was found embedded in Mesozoic Era sediments. This coiled cephalopod was ancient when dinosaurs walked the earth and a dinosaur may well have stepped on this specimen some 70 million years ago. The chunk of rock containing this fossil was deposited somehow, 75 million years ago, in what is thought to have been a Late Mesozoic watering hole; at that time the cephalopod was already 420 million years old. Chronister Vertebrate site, Bollinger Co., Mo.

Straight nautaloid cephalopods from the Canadian (younger Lower Ordovician).

Fig. 03-46. *Cotteroceras compressum,* Ulrich and Foerste. These straight cephalopods are quite early (as straight cephalopods go). Straight cephalopods are a dominant Paleozoic mollusk which span most of the Paleozoic Era and then go extinct at the end of the Permian Period. These dolomite specimens came from a layer in a road cut. When the rock cut was removed to accommodate yuppie gentrification, the fossiliferous rock was carted away and gone through for its fossils at a later time, an excellent collecting strategy for endangered fossil horizons. Cotter Formation, Crystal City, Mo. (Value range G, single specimen).

Fig. 03-47. *Cotteroceras compressum.* Another group of straight cephalopods from a different locality, this time in Arkansas. Early straight cephalopods like these were often gregarious so that groups of them are found parallel to each other, sometimes they are stacked like cordwood. Cotter Formation, Yellville, Arkansas.

Fig. 03-48. *Protocycloceras* sp. A species of this straight cephalopod which has a shell surface exhibiting distinctive ornamentation. Such ornamented cephalopods are more characteristic of the Middle Ordovician or later in geologic time than they are this early. Cotter Formation, Jefferson County, Missouri. (Value range G, single specimen).

Coiled cephalopods from the late or younger Lower Ordovician (Canadian).

Fig. 03-49. This nice specimen of an early coiled nautaloid is from the Cotter Formation. The particular specimen is preserved in dolomite and was associated, as usual, with stromatolites. Usually such fossil cephalopods are preserved in chert. Cotter Formation, Crystal City, Missouri. (Value range E).

Fig. 03-50. These small coiled nautaloid cephalopods can occur in some abundance, usually associated with stromatolite reefs upon which the nautaloids presumably fed. Such an obvious association with stromatolites is absent after the Ozarkian and Canadian and it is believed that as biomass, particularly that of browsers, increased, stromatolite reefs were cropped as soon as they could form. Thus ended an ecosystem (with some exceptions) which had its beginnings over three billion years earlier. Cotter Formation, Pontiac, Missouri.

Fig. 03-51. *Campbelloceras* sp. A superb specimen of this early coiled nautaloid preserved in chert. Such large, well preserved nautaloids are rare and are therefore desirable. Nautaloid cephalopods, from which the group gets its name, are represented in today's oceans by the pearly nautilus, Smithville, Formation, Bollinger Co., Missouri. (Value range E).

Fig. 03-52. *Centrotarphyceras yellvillensis.* A group of large coiled nautaloids from a prolific site for them near Mountain Home, Arkansas. Note that the whorl becomes separated from the shell's coil in the vicinity of the living chamber. This is an example of a disjunct coiled nautaloid, a characteristic of many Lower Ordovician (Canadian) coiled cephalopods. Unfortunately the locality where these were found has been affected by urbanization and material is now difficult to obtain. Powell Formation, Tracy's Ferry, Mt. Home, Arkansas. (Value range E, single specimen).

Fig. 03-53. *Centrotarphyceras yellvillensis*, with an illustration of a similar specimen from late 19th century paleontologic literature. Ozarkian and Canadian fossils remained relatively unknown until the mid-20th century when focus on the Ozarkian and Canadian periods increased scientific interest in them, The Ozark Uplift, the most prolific area for them, remained in some ways geologically unknown until the 1920s and 1930s.

Fig. 03-54. A group of fine, silicified nautaloids from the Upper Canadian Smithville Formation of Bollinger County, Missouri. Bottom row *Cycloplectoceras fuwatum*. Top row left: *Cyclopstomiceras* sp.; right: *Wichitoceras compressum*. Specimens identified by the late Rousseu Flower.

Fig. 03-55. *Campbellloceras virginianum*, Hyatt. A silicified specimen which includes the silicified nautaloids shell so that chambers, normally seen on chert steinkerns (internal molds) are not visible. Smithville Formation, Bollinger County, Mo. (Value range F).

Fig. 03-56. *Lituites lituus*. These highly disjunct cephalopods (whorl becomes separated at outer part of shell) come from the Isle of Oland, Sweden, an island off of the mainland of Sweden in the North Sea. These cephalopods occur in a distinctive red limestone of Canadian age. The Oland occurrence is unique to Europe. *Lutuites* of this type are not found in other strata of the same age in Europe or in North America. (Value range G, single specimen).

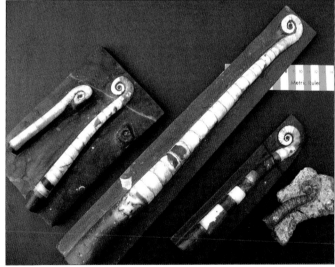

Fig. 03-57. Trilacinoceras hunanense, La et Tsi. These disjunct cephalopods from China, came onto the fossil market in quantity in 2001. They are virtually identical to Lituites lituus from Oland, Sweden (see above). They also occur in the same red limestone as the Swedish specimens. The Cambrian of China can be subdivided into four distinct provinces, each one possibly a fragment of a Cambrian continent that was "welded" together sometime during a later part of geologic time. Lower Ordovician paleogeography is quite similar to that of the Cambrian and these two occurrences of Lutuites, one in Sweden, the other in China, on opposite sides of the globe, may represent two pieces of the same landmass now separated by sea floor spreading and continental drift. (Value range F, single specimen).

Trilobites

Lower Ordovician trilobites which lived on the craton are generally found incomplete like those of the Upper Cambrian. There is also much species diversity similar to that of the late Cambrian.

Fig. 03-61. *Jeffersonia* sp. A group of early Canadian trilobite partials. Jefferson City Formation, Phelps Co., Missouri. (Value range F, group of specimens).

Fig. 03-58. *Hystricurus missourensis*. A distinctive and highly pustulose trilobite associated with shallow seas of cratonic North America. Ozarkian and Canadian trilobites of the craton are usually different from those which lived in areas of deeper water. Such trilobites are found in dark colored, slaty rocks formed at the margins of Cambrian and Lower Ordovician continents and are quite different looking from these specimens preserved in yellow chert. Gasconade Formation, Crawford Co., Missouri (Value range E, for group).

Fig. 03-59. *Hystricurus missourensis*. Another group of these bumpy or pustulose trilobite partials. Gasconade Formation, Lower Ozarkian.

Fig. 03-62. Cranidia of typical early Canadian trilobites characteristic of the craton of North America. Lower Ordovician trilobites like these, are quite different from those that lived in the deeper waters of the edge of the continent or in the open ocean. Jefferson City Formation, Phelps County, Missouri.

Fig. 03-60. *Paraplethopeltis* sp., Bridge and Cloud, 1942. An early most Ordovician trilobite cephalon from a genus similar to the Cambrian genus *Plethopeltis*. The authors of the species, J. Bridge and Preston Cloud were both advocates for Ulrich's Ozarkian and Canadian Periods. Bridge coauthored many Ozark fossils as well as doing fundamental geologic mapping and exploration in the Ozarks during the 1920s and 1930s. Preston Cloud later became a world authority and leader on the Precambrian and its fossils in the 1960s and 1970s. Gasconade Formation, Carter Co., Missouri.

These trilobites of the Atlantic faunal province are from geosynclinal sediments deposited in moderately deep waters. In the deeper waters of a geosyncline, Lower Ordovician faunas are different from those of the craton with its stromatolites. These deeper water faunas are in some ways more attuned to the Cambrian, especially the trilobites; Middle and Late Ordovician trilobites by contrast are quite different. Sediments deposited in the open ocean will contain fossil graptolites, colonial hemichordates whose pelagic (floating) colonies sank and were buried by ocean bottom sediments. Through sea floor spreading, such deep sea sediments were then "plastered" on to the margins of continents, where the sediments were crumpled and often metamorphosed.

Fig. 03-63. *Niobella (Asaphellus) homfrayi*. This trilobite comes from Tremadocian strata which is considered by British stratigraphers to represent the uppermost beds of the Cambrian System. Others consider the Tremadocian as the earliest (lowermost) strata of the Ordovician. It is quite difficult to correlate trilobite and graptolite bearing strata of Europe (particularly Wales) with that of the molluscan-rich Ozarkian strata of North America. Very few fossils are common to these two sequences, each being deposited under very different marine conditions, one in shallow, stromatolite dominated waters, the other in deep waters. Avalon Peninsula, Newfoundland. (Value range F).

Fig. 03-64. *Asaphellus (Niobella) homfrayi*. This trilobite from the Tremadoc Series of Wales is identical to the previously shown specimens from Newfoundland. They both lived in the same body of medium deep seas whose sea floor sediments are now separated by the Atlantic Ocean, a relatively modern feature of the earth's crust which formed much later in geologic time than the Cambrian or the Tremadocian. Tremadoc Series, Shineton, Salup, Wales. (Value range F).

Fig. 03-67. *Ogygiocarella debuchi*, Brongniart. A group of these representative "deep sea" trilobites from the Landeilo Series of Wales.

Fig. 03-65. *Asaphellus homfrayi*. Two distorted specimens of this dominant Tremadocian trilobite from Wales. The Welsh section of strata which yields these fossils has been folded and mildly metamorphosed, resulting in distortion of the trilobites as shown here. Distorted trilobites (and other fossils) have sometimes been described as "new" species or genera when they are the same forms that are found in non-distorted strata of the same age.

Fig. 03-68. *Cremidopyge bisecta*, Murchison, 1839. An early representative of a family of trilobites which are more representative of the Middle and Upper Ordovician or even the Silurian Period than of the Lower Ordovician. This trilobite comes from Llandelian age strata of Wales. The Llandelian is placed in the Lower Ordovician by some geologists and in the Middle Ordovician by others. It is difficult to correlate deep sea Ordovician strata, like that of Wales, with that of the craton, as these two different environments have few fossils in common. (Value range F).

Fig. 03-66. *Ogygiocarella debuchii*, Brongniart. This trilobite is from the late Lower Ordovician or early middle Ordovician, Llandeilo Series of Wales. It is unclear where the Lower-Middle Ordovician boundary occurs in the thick, slaty strata of Wales that yield these trilobites. Many of the Lower Ordovicain trilobites of Wales are part of an edemic "Atlantic Province" fauna and many, like these, still have a Cambrian aspect about them. Radnorshire, Wales (Value range E).

Fig. 03-69. *Asaphus expansus.* A robust trilobite from the Isle of Gotland in Sweden. *Expansus* limestone, Ljungsbro, Sweden. This specimen came through Geological Enterprises, a dealer in fossil specimens in Ardmore, Oklahoma. (Value range F).

These trilobites came from various faunal provinces of the world.

Fig. 03-70. *Hoekaspius (Megalaspis) matalaspis*, Hoek. These relatively large trilobites occur in hard, flinty concretions which weather from thick beds of black, slaty shale exposed in the Andes Mountains of Bolivia and Argentina. A number of these came onto the fossil market around 1980. They are Canadian = Llanvirnian in age; that is late Lower Ordovician. San Lucas Ocuri Formation. Nor Cinti near San Lucas, Chuquisaca Province, Bolivia. (Value range E).

Fig. 03-71. *Hoekaspius matalaspis*, Hoek. A concretion with part and counterpart of this trilobite from Bolivia. Late Lower Ordovician (High Canadian). (Value range F).

Fig. 03-72. *Asaphopsis* sp. A species of this Lower Ordovician trilobite with two pronounced spines on the pygidium. Hunan Province, China.

Fig. 03-73. *Norinia convexa*. The Cambrian (and Lower Ordovician) of China consists of a number of faunal provinces (often considered to be four) which represent pieces of continental crust which were "welded" together sometime during later geologic time to produce part of Asia. One of these faunal provinces is similar to the Cambrian of California and Nevada. These trilobites come from a black limestone of possibly lowermost Ordovician (Tremodocian) age whose trilobite fauna is somewhat similar to that of Newfoundland and Wales.

Fig. 03-76. *Phyllograptus* sp. Part and counterpart of the same slab as above. Ibex area, House Range, Utah.

Fig. 03-74. *Norina convexa*. A large specimen of this nice trilobite from the Lower Ordovician of China. (Value range E).

Graptolites

Fossils of these colonial and pelagic (floating) chordates are typical of Ordovician deep sea sediments.

Fig. 03-77. *Phyllograptus elongatus* and *Tetragraptus bigsbyi,* Hall. These graptolites are typical of graptolite preservation, a shiny film of graphite on black, slaty shale. From the deep sea, graptolite facies of the Lower Ordovician. *Didymograptus* shale, Oslo, Norway. (Value range G).

Fig. 03-75. *Phyllograptus* sp. A distinctive Lower Ordovician graptolite in which the theca has been preserved as a brown, organic residue rather than as the usual carbon or graphite film on black, slaty shale. The graptolite theca was originally composed of a protein similar to that which composes your fingernail. This specimen came from a sequence of Ozarkian and Canadian strata formed near the edge of the North American craton. Such strata was never deeply buried or metamorphosed, as is often the case with other graptolite bearing strata. This area, the Ibex area of western Utah, has recently become the type region for Lower Ordovician strata and biostratigraphy in North America. Fillmore Formation, Ibex area, House Range, Millard Co., Utah. (Value range G).

Fig. 03-81. *Phyllograptus angustifolius.* Tremadocian, Slemmestad, Norway. Specimen from Malicks' Fossils, a scientific fossil dealer from the 1940s through the 1960s.

Fig. 03-78. *Phyllograptus* sp. A characteristic Lower Ordovician graptolite preserved as a graphite film in black, slaty rock. Phi Kappa Formation, Trail Creek, Idaho.

Fig. 03-79. *Tetragraptus similis* Phi Kappa Formation, Trail Creek, Idaho. (Value range H).

Fig. 03-82. *Phyllograptus angushfolius.* From deep sea, black shales of Norway. Such deep sea sediments were "scraped" from the sea floor by sea floor spreading and piled against both North America and Europe. Tremadocian, Slemmestad, Norway.

Bibliography

Ulrich, E. O. 1911. "Revision of the Paleozoic Systems" *Bulletin of the Geological Society of America.* Volume 22, pp. 281-680.

Ulrich, E.O., A.F. Foerste, and A.K. Miller. "Ozarkian and Canadian Cephalopods, Part II: Brevicones," Geological Society of America, Special Paper No. 49.1943,

Ulrich, E.O., A.F. Foerste, A.K. Miller, and W. M. Furnish. Ozarkian and Canadian Cephalopoda. Part I; Nautilicones. Geological Society of America Special Paper No. 37, 1942.

Ulrich, E.O., and A.G. Unklesbay. "Ozarkian and Canadian Cephalopods, Part III. Longicones and Summary." Geological Society of America, Special Paper No. 58. 1944.

Fig. 03-80. *Loganograptus* sp. Phi Kappa Formation, Trail Creek, Idaho. (Value range H).

Chapter Four

The Ordovician Period

Middle and Upper Ordovician

Receptaculitids.

They are probably a calcareous red algae, but are commonly (and incorrectly) classified as a sponge.

Fig. 04-01. *Receptaculites oweni*, Hall. This is a common fossil in Middle and Upper Ordovician rocks of the continental interior of North America. When *Receptaculites* was described in the 19th century, it was considered to be either a sponge or a coral. It is now considered by most paleontologists as a type of red or dasycladacian algae. The species name *oweni* was named after pioneering geologist David D. Owen who first recognized it, although the generic name *Receptaculites* was given to it by James Hall. Hall's name became well established, however, in 1979 for legalistic reasons seemingly in violation of ICZN decisions, the name was changed to *Fisherites* (Finney and Nitecki, 1979). It is the author's opinion that such a well established name as *Receptaculites* should remain and in light of the fact that this is a non-refereed publication, the author retains the use of Hall's well established genus. The specimen shown here has more depth than is usual and is preserved in chert. Kimmswick Formation, Middle Ordovician, Jefferson County, Missouri. (Value range F).

Fig. 04-02. *Receptaculites oweni*, Hall. Another relatively small, "3-D" specimen of *Receptaculites* from Middle Ordovician limestone. Recently, as a consequence of strict application of taxonomic legalities (see above), the well established genus *Receptaculites* is now "officially" *Fisherites* sp. Glen Park, Jefferson County, Missouri. (Value range F).

Fig. 04-03. *Receptaculites oweni*, Hall. A typical flat form of this calcareous red algae. *Receptaculites* occurs over a large part of cratonic North America. This specimen is from Upper Ordovician strata just west of the Canadian Shield in northern Saskatchewan, Canada. Most Receptaculitids come from Iowa, Missouri, or Illinois. (Value range F).

Fig. 04-04. A weathered section through a small *Receptaculites* specimen. Middle Ordovician, Kimmswick Limestone, Jefferson County, Missouri. (Value range F).

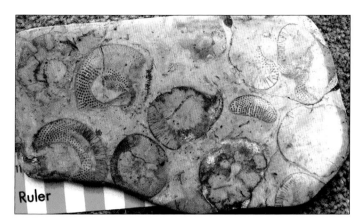

Fig. 04-05. *Ischadites iowensis,* Owen. A polished slab with a group of these small Receptaculitids. *Ischadites* is frequently found clustered together as on this slab and it usually is not associated with specimens of *Receptaculites.* Galena Formation, Cannon Falls, Minnesota. (Value range F).

Fig. 04-08. Castellated bluffs of Middle Ordovician limestone exposed on the Iowa River in northeastern Iowa. From David D. Owen, "Report of the Geological Survey of Wisconsin, Iowa and Minnesota," 1852.

Sponges

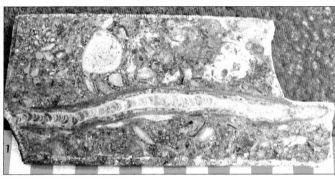

Fig. 04-06. *Ischadites iowensis,* Owen. A group of these small receptaculitids from Middle Ordovician strata of northeastern Iowa. *Ischadites* is usually found separate from larger specimens of *Receptaculites oweni.* Middle Ordovician, Prossert Limestone, northeastern Iowa. (Value range G, single specimen).

Fig. 04-09. *Stromatocerium (Labechia) rugosum,* Hall. This is a peculiar type of sponge known as a sclerosponge. Sclerosponges were thought to have become extinct after the Paleozoic Era, however living specimens of this lower invertebrate were found in the 1980s in submarine caves of Jamaica, W. I. (Value range G).

Fig. 04-07. Upper Ordovician limestone exposed in the Big Horn Mountains of northern Wyoming.

Fig. 04-10. Sponges. These elongate sponges occur on slabs of shale, where they are often clustered together. The shale layers are separated by a peculiar cephalopod rich coquina, specimens of which have been widely distributed, but these associated sponges are less so. Maquoketa Formation, Graff, Iowa. (Value range F).

Cnidarians

The phylum to which jellyfish and coral belong.

Fig. 04-11. *Chonchopeltis* sp. *Chonchopeltis* is believed to be a Chrondrophorine cnidarian, a relative of the "by-the wind-sailor" or the Portuguese-man-o-war. They have a leather-like cap and it is this cap which made this impression. Chrondrophores are relative rare fossils and impressions such as these were previously placed as a type of early mollusk. Ellis Yochelson, in the 1990s, sorted out many of the different circular medusoid fossils and determined that this type of lower Paleozoic "medusoid" is the bell of a chrondrophore cnidarian. Decorah Formation, Lincoln County, Missouri. (Value range G).

Fig. 04-13. *Columnaria halli*. This is one of the earliest, commonly found fossil corals. Although corals are found in rock strata older than the Middle Ordovician, earlier ones are not usually conspicuous. *Columnaria* is the oldest coral which casual fossil collectors usually take notice of. The specimens have been silicified (replaced with quartz) and have been released from the rock by weathering of limestone beds. Plattin Formation, Middle Ordovician, Jefferson County, Missouri. (Value range F, single specimen).

Fig. 04-12. *Conularia* sp. Conularids are problematic fossils that some paleontologists (but not all) consider to be related to Chrondrophore cnidarians (see previous image). Their ornamented "shell" is not mineralized but rather appears to have been composed of some sort of polysaccharide compound such as chitin (the composition of insect exoskeletons) or possibly cellulose. Maquoketa Formation, Upper Ordovician, Lincoln County, Missouri. (Value range F).

Fig. 04-14. *Columnaria halli*. A silicified specimen of this early coral. The geologic association of this coral occurrence is unusual, it was found "out of place" in strata associated with late Mesozoic dinosaur bearing sediments of the Chronister vertebrate site, Bollinger County, Mo. (Value range F).

Bryozoans

Very common fossils in Middle and Upper Ordovician marine limestones, deposited on the craton.

Fig. 04-15. *Hallopora rugosa* (Bryozoan). A nice branching colony of bryozoans from Upper Ordovician strata of Kentucky. The species name refers to the rugose pattern characteristic of this species. Bryozoans are not usually readily identified as megafossils; their identification generally requires microscopic examination of the zooecia in thin section. From thinly bedded, Upper Ordovician limestones which underlie central Kentucky (Bluegrass region). (Value range F).

Echinoderms

These exclusively marine animals become abundant in the Middle Ordovician and remained dominant in the oceans until the present. Their best known representatives are sea urchins and starfish.

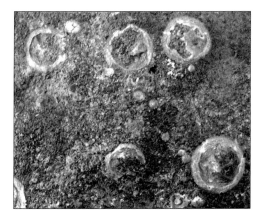

Fig. 04-17. *Echinospherites aurantium.* This is one type of an extinct echinoderm class known as the cystoid. *Echinospherites* occurs as an almost perfect sphere and may have been transported when living by rolling with wave action on the shallow sea floor. Bottom two specimens from the Kimmswick Limestone, southeastern Missouri. Top specimen from the Benbolt Limestone, Middle Ordovician, Scott Co., Virginia. (Value range F).

Fig. 04-18. *Castericystis vali.* Pleurocystids (*Pleurocystites* and the related *Castericystis*) are widespread cystoid genera in the eastern half of North America. These specimens come from quarries in eastern Ontario working Middle Ordovician limestones that occur as flat lying beds sandwiched between the Precambrian Laurentian mountains to the north and the folded Appalachian mountains to the south. Verulam Formation, Middle Ordovician, Simcoe Co., Ontario, Canada. (Value range E, single slab).

Fig. 04-16. *Isorophusella incondite.* These disk-shaped echinoderms are edrioasteriods, one of the extinct echinoderm phyla of the Paleozoic. Edrioasteriods often were attached to "hard grounds," hard, lime mud surfaces on the sea floor, or to the surfaces of shells such as those of large brachiopods. Verulan Formation, Middle Ordovician, Simcoe County, Ontario.

66

Fig. 04-19. *Pleurocystites.* *Pleurocystites* is a cystoid, an extinct Paleozoic echinoderm class. In the 1980s a number of echinoderm classes were established from various types of early Paleozoic cystoids. The class cystoidea had become somewhat of a "catch all" class for Paleozoic echinoderms that didn't seem to fit into other well defined echinoderm classes, such as the blastoids and crinoids. Middle Ordovician, Ralls County, Missouri. (Value range G, single specimen).

Fig. 04-20. *Pleurosystites* sp. Another selection of these interesting and desirable cystoids, *Pleurocystites* represents a cystoid particularly characteristic of the Middle and Upper Ordovician. Bottom specimen, Upper Ordovician, Maquoketa Formation, northeastern Iowa, top, Middle Ordovician, southeastern Minnesota.

Fig. 04-21. *Scalenocystites strimplei,* Kolata. These have been placed in the extinct echinoderm class Stylophora on the basis of their distinctive stele (tail) as well as other unique, anatomical features. Some paleontologists consider stylophorans as representing a link between chordates (vertebrates) and echinoderms and have established a separate extinct phylum for them, the Calcichordata. *Scalenocystites,* before the 1980s, was considered to be a type of cystoid. Galena Formation, Middle Ordovician, Cannon Falls, Minnesota. (Value range E.)

Fig. 04-22. *Scalenocystites strimplei,* Kolata. These fossils represent one of the echinoderm classes established from what were previously classified as a type of cystoid. The species name is for Harold Strimple, a devoted worker on Paleozoic crinoids at the University of Iowa from the 1940s through the 1970s.

Fig. 04-23. *Cremacrinus articulosos,* Billings. These peculiar crinoids, known as "nodding crinoids," occur on the surfaces of slabby limestones (bedding planes) of middle and upper Ordovician age in eastern North America. They are desirable crinoids. Decorah Formation, Jefferson Co., Missouri (Value range E).

Fig. 04-24. Starfish; *Taeniaster* sp. Impressions of these small starfish have become widely distributed among collectors as they are associated with a well collected zone of Cryptolithid trilobites. The starfish impression is naturally accented with iron oxide (limonite). Martinsburg Formation, Swatara Gap, Pennsylvania (Value range F).

Fig. 04-25. *Phragmactis* sp. A small starfish, similar to that of the previous image, associated with greywacke, a type of dirty sandstone often indicative of a relatively deep sea. These starfish probably lived on the floor of a relative deep sea in a manner similar to starfish found on today's deep ocean floor. Middle Ordovician, Scotland. (Value range G).

Fig. 04-26. Ophuroid. These impressions of brittle stars are associated with what appear to be deep sea crinoids and appear to be part of a relatively deep sea fauna. Associated with these echinoderms in the same series of strata are specimens of *Eldonia*, a problematic vendozoan-like fossil known also from the Middle Cambrian Burgess Shale of British Columbia. Ordovician, Atlas Mountains, Morocco. (Value range E).

Fig. 04-27. Ophuroid. Another part of the same slab as above. The elongate fossils are crinoids; all are preserved as iron stained impressions in "dirty" sandstone known as greywacke, which is often indicative of sediments deposited in part of a deep sea trench. Middle (?) Ordovician, Atlas Mountains, Morocco.

Fig. 04-28. Close-up of one of the ophuroids, shown in the previous two images.

Mollusks

Mollusks are abundant fossils in mid and late Ordovician rocks particularly in that strata deposited on the craton.

Fig. 04-29. *Polylopia* sp. An example of a problematic post-Cambrian mollusk. These fossils were originally considered as early scaphopods, a molluscan class which is well represented in shallow waters of today's oceans. The shells of *Polylopia* telescope in a peculiar manner unlike scaphopods and no other scaphopod-like fossils are known this early in the fossil record. Ellis Yochelson, considers *Polylopia* as representing an extinct molluscan class. Murfreesboro Limestone, Murfreesboro, Tennessee. (Value range F).

Fig. 04-30. *Maclurites logani*, Salter. The operculum or "cap" of a large, sinistral coiling snail characteristic of the Middle and Late Ordovician. Plattin Limestone, Middle Ordovician, Jefferson County, Missouri. (Value range F).

Fig. 04-31. *Maclurites logani*, Salter. These large, left-hand-coiling gastropods can be common in Middle and Late Ordovician limestones of the North American craton, that part of the continent covered by shallow seas during the Paleozoic Era. The fossil illustrated in the previous image is believed to be the operculum of the animal which occupied this shell. *Maclurites* is usually found and preserved as internal fillings of its large shell. Middle Ordovician, northeast Iowa. (Value range G).

70

Fig. 04-32. *Crytolites* sp. The bellerophonts are planespirally coiled gastropods of the Paleozoic Era (and also the Triassic Period of the Mesozoic Era). Some paleontologists consider bellerophonts as a type of monoplacophoran and not a gastropod. Kimmswick Limestone, Middle Ordovician, Franklin County, Missouri. (Value range F).

Fig. 04-33. "*Orthoceras*" or *Endoceras*. These large, straight cephalopods occur with some abundance in the Middle and Upper Ordovician. The length of the complete animal including the head which extended beyond the living chamber, would have been between two and three meters long, a pretty large mollusk. Portions of these large, straight cephalopods can be fairly common. Complete ones or reasonably complete ones are less so. Plattin Limestone, Middle Ordovician, Jefferson County, Missouri.

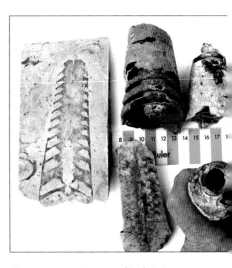

Fig. 04-35. *Actinoceras bigsbyi*. A group of robust straight cephalopods with a complex and large siphuncle. Plattin Limestone, Middle Ordovician, Jefferson Co., Mo.

Fig. 04-34. *Endoceras proetiforme*. These straight cephalopods occur in a concentration, sometimes with one telescoped within another. They can be common fossils but like many (perhaps most) fossils, they don't look like much until prepared and that can take a considerable amount of time. Maquoketa Formation, Graff, Iowa. (Value range G).

Fig. 04-36. *Gonioceras ancepts*. A flat, triangular shaped cephalopod which has been suggested to have occupied the same ecological niche as that occupied by sting rays today. *Gonioceras* is a characteristic Middle Ordovician index or guide fossil. Platteville Formation, northern Illinois. (Value range F, single specimen).

Fig. 04-37. "Cigaroceras"! No, that's not the actual genus. This cephalopod is one of many different cephalopod shapes that appear after the initial appearance of cephalopods in the Lower Ordovician, some take on crazy shapes. Plattin Limestone, Middle Ordovician. (Value range F).

Arthropods

Arthropods are segmented, multi-legged invertebrates. Here are two non-trilobite arthropod fossils.

Fig. 04-38. *Leperditia fabulites*, Conrad. Ostracodes are a type of Crustacean, in the same arthropod class as crabs, lobsters, and shrimp. They are informally known as clam-shrimp. Normally they are microfossils, these, however, are giants for ostracodes. The Middle Ordovician was a time of giant species and genera of ostracodes. Stones River Formation, Middle Ordovician, Murfreesboro, Tennessee.

Fig. 04-39. *Leperditia* sp. Ostracodes are normally thought of as being microfossils. In the Ordovician giant representatives of these "clam shrimp" were locally abundant. These are some of the smaller of the giants. Dutchtown Formation, Lower Middle Ordovician (Chazy Series). Cape Girardeau County, Mo. (Value range F).

Trilobites

Trilobites such as these are found in limestones of Middle and Upper Ordovician age deposited on the craton.

Fig. 04-40. *Ceraurus pleurexanthemus*, Green. An interesting and distinctive trilobite characteristic of the Middle Ordovician of North America which when found as a complete specimen, is quite desirable. Decorah Formation, Middle Ordovician, Lincoln County, Missouri. (Value range E).

Fig. 04-42. *Isotelus gigas. Isotelus* is a common trilobite of the Ordovician Period. In the late 19th century and early 20th century, black limestones of Trenton Gorge, in upstate New York, produced a number of these trilobites that, when seen today, come primarily from old collections. A number of pioneering North American paleontologists, including James Hall and Charles D. Walcott, "cut their teeth" at fossil collecting at the Trenton Falls locality. (Value range E).

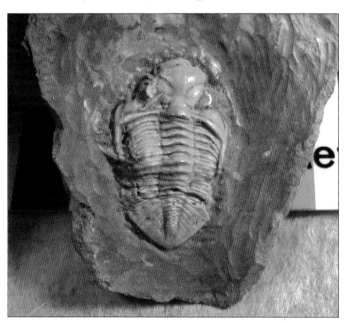

Fig. 04-41. *Eomonarachus* sp. A small but quite nice trilobite characteristic of the Middle Ordovician of the craton of North America. Decorah Formation, Silex, Missouri. (Value range F).

Fig. 04-43. *Bumastus* sp. The pygidia of this ubiquitous Middle Ordovician trilobite genus. Trenton Falls, New York.

Fig. 04-44. *Isotelus* sp, Hyperstome. This structure, at the bottom front (front ventral) part of a trilobite is the hyperstome. Those of the Ordovician genus *Isotelus* are particularly distinctive. Middle Ordovician, Nashville, Tennessee. (Value range H).

Characteristic Ordovician trilobites that lived in the deeper waters of a geosyncline or on the margin of a depressed continent.

Fig. 04-45. *Asaphus* sp. Trilobites such as this come from folded Ordovician shales of the mountains of Spain and southern France. This specimen from Spain, shows the impression of the hyperstome of the trilobite, an anatomical structure unique to trilobites and possible involved with feeding. Middle Ordovician, Caceres, Spain. (Value range F).

Fig. 04-46. *Illaenus hispanicus*. An enrolled specimen of this widespread Ordovician trilobite genus. Trilobites of the Ordovician of Spain and southern France (Pyrenees Mts.) are similar in many ways to those of Morocco. Middle Ordovician, Caceres, Spain.

Fig. 04-47. *Illaenus hispanicus*. Another enrolled specimen. Post Cambrian trilobites could enroll in this manner as a defense mechanism when threatened, to protect the soft appendages on the animals ventral (bottom). Middle Ordovician, Caceres, Spain. (Value range F).

Fig. 04-48. *Asaphus brachycephalus*. This large, iron- stained Moroccan trilobite is a member of the Family Asaphidea. The genus *Asaphus* is one of the most common of the large trilobites found in Europe and Morocco. In North America, most large Asaphid trilobites are assigned to the genus *Isotelus*. This *Asaphus-Isotelus* nomenclatural duo evolved before the advent of plate tectonics and was often stubbornly used by 20th century paleontologists who refused to accept the concept of continental drift. Essentially species of *Asaphus* and *Isotelus* converge on each other, If this trilobite were found in North America it would be in the genus *Isotelus*. Moroccan trilobite nomenclature follows that of Europe so that this specimen belongs to the genus *Asaphus*. Middle Ordovician, Atlas Mts., Morocco. (Value range F).

Fig. 04-49. *Asaphus brachycephalus.* A large specimen of *Asaphus* preserved in a ferruginous concretion. Numerous large Ordovician trilobites from the Atlas Mountains of Morocco have come onto the fossil market in the early part of the 21ˢᵗ century at very low prices, particularly considering the size of the trilobites. (With trilobites, beside being best if complete, the bigger the trilobite the better!) Rumors have circulated that these rusty looking specimens are not real fossils but are replicas. The author has not found this to be the case, but rather these specimens represent an opportunity to obtain some fantastic trilobites, that are usually seen only in museums (if then), for a very low price. Middle Ordovican, Atlas Mountains, Morocco. (Value range E).

Fig. 04-51. *Scutellum* sp. A complete specimen (top, right) and two pygidia of this trilobite. Kimmswick Limestone, Middle Ordovician, Glen Park. Missouri.

Fig. 04-50. *Calymene tristani.* Trilobites of the genus *Calymene* range from the Middle Ordovician through the Silurian and into the Lower Devonian. This elongate specimen, is from the Ordovician of Spain. Cacares, Spain. (Value range G).

Fig. 04-52. *Illaenus americanus.* The genus *Illaenus* is a wide-spread Ordovician trilobite both in North America and Europe. Specimens are usually found in dirty sandstone or in siltstone such as the specimens from Morocco; specimens in limestone like this one are relatively rare. Kimmswick Limestone, Ralls County, Missouri. (Value range E).

Fig. 04-53. *Amphilichus cuculas,* Bradley. The specimen on the left preserves, in three dimensions, the original pustulose carapace of these peculiar trilobites. The specimen on the right has partial original material. Such 3-D partial trilobites from the U.S. midwest are undervalued as they often are well preserved and preserve the three-dimensional shape of the cephalon; they also are not easy to collect. Middle Ordovician, Kimmswick Limestone. Jefferson County, Missouri. (Value range G).

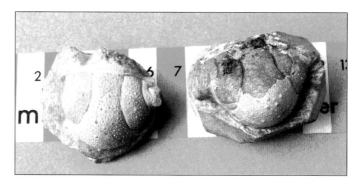

Fig. 04-54. *Homotelus bromidensis*, Esker. A concentration of these trilobites occurs in the Arbuckle Mountains of Oklahoma, where many specimens, often on slabs like this, have been distributed worldwide by Geological Enterprises of Ardmore, Oklahoma. The author of the species, George Esker, working with Dorothy Echols of Washington University in the early 1960s, put his namesake on this well known and showy trilobite. Bromide Formation, Middle Ordovician, Arbuckle Mountains, Oklahoma. (Value range C).

Fig. 04-55. *Homotelus bromidensis*, Esker. Slightly larger specimens than those of the previous image, from the Bromide Formation of the Arbuckle Mountains, Oklahoma. (Value range C).

Fig. 04-57. *Illaenus* sp. Numerous complete specimens of this Ordovician genus have come from the Atlas Mountains of Morocco. Often they are labeled as being from Eufrod, Morocco, however, this is a small city near the mountains where many wholesale fossil dealers are located and is not the actual locality from which the fossils originated. The localities for Moroccan trilobites are numerous and are scattered throughout the Atlas Mountains.

Fig. 04-56. *Illaenus* sp. An example of a flat specimen of essentially the same trilobite as previously illustrated from Spain. Like the Spanish trilobites, these from Morocco are preserved as internal molds in fine grained greywacke (dirty sandstone). Atlas Mountains, Morocco. (Value range G).

Fig. 04-58. *Isotelus iowensis*, Owen. Complete specimen but with a damaged pygidium. Maquoketa Formation, Turkey River, Iowa. Ordovician trilobites were first noted along the Turkey River in the northeastern part of Iowa Territory by David D. Owen in the 1830s. Owen was a pioneer geologist of the U.S. midwest and the author of this species of *Isotelus*. Slabby limestones of the Middle and Upper Ordovician crop out along the Turkey River and produce nice fossils. (Value range F).

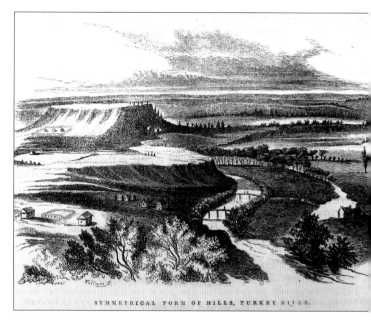

SYMMETRICAL FORM OF HILLS, TURKEY RIVER.

Fig. 04-60. View of the Turkey River locality as it appeared circa 1835. Upper Ordovician, slabby limestone crops out along the edge of the river. The mound or butte-like hill in the background is composed of overlying Silurian Dolomite which yields the peculiar receptaculitid *Cyclocrinites* sp. (see Silurian chapter). David D. Owen traversed, by canoe, this and all of the other small rivers in Iowa and Minnesota Territories flowing eastward into the Mississippi in the 1830s and 1840s. He mapped the geology and rocks exposed along the rivers for their mineral potential. If extensive mineralization potential was found, such areas were to be set aside as mineral lands which were not opened for homesteading when statehood arrived. From David D. Owen, 1853.

Fig. 04-59. *Isotelus iowensis*, Owen. Maquoketa Formation, Upper Ordovician, Turkey River, northeastern Iowa. This specimen is a trilobite molt. Trilobites, like crabs and cicadas, would periodically shed their exoskeleton to allow for growth of the animal inside. The skewed angle of the trilobites glabella in relationship to the thoraxial segments is indicative that this specimen is a molt. (Value range G, for molt).

Fig. 04-62. *Isotelus gigas,* Dekay. Two small specimens of this Middle Ordovician species of *Isotelus* collected from the Salt River, one of the small rivers that flow into the Mississippi in northeastern Missouri, and one also evaluated by David D. Owen. Decorah Formation, Ralls County, Missouri. (Value range E).

Fig. 04-63. *Selenopeltis buchi,* Hawle and Corda. These large trilobites come from a zone rich in them and discovered in the 1980s in the Atlas Mountains of Morocco. In this zone specimens of this normally rare trilobite are clustered together as can be seen on this slab. *Selenopeltis* is also found in Europe from where it was first described, but it is a relatively rare genus. Upper Ordovician, Borji, zona di Erfoud, Morocco. (Value range B).

Opposite
Fig. 04-61. *Isotelus iowensis,* Owen. Another Turkey River *Isotelus* from limestone ledges exposed along the Turkey River of northeast Iowa (see previous image). Maquoketa, Formation, Clermont, Iowa. (Value range F).

Fig. 04-64. *Flexicalymene meeki,* Foerste. These early calymenid trilobites are relatively abundant in Upper Ordovician shales and shaley limestone around the Cincinnati, Ohio, area and parts of southern Ohio. An abundance of fossils in the Cincinnati area has encouraged a number of its residents to take up paleontology, including August Foerste, the author of the species. (Value range G).

78

Fig. 04-65. *Ogygites canadensis.*
Ogygites occurs locally in abundance
in black, petroliferous shale at
Georgian Bay, near Collingwood,
Ontario. Collingwood Shale. (Value
range F).

Fig. 04-66. *Triarthrus beckeri.* This is a common trilobite in
black, petroliferous shale of the Upper Ordovician Utica
Formation of upstate New York and adjacent Ontario, Canada.
Specimens of *Triarthrus* preserving the appendages in pyrite
were found in the mid 19th century by Charles D. Walcott.
(Value range H).

Fig. 04-67. *Calymene tristani.* This
Ordovician species of *Calymene*
was one of the first species of this
widespread trilobite genus *Calymene*
to be described. Middle Ordovician,
Cacares, Spain.

Fig. 04-68. *Neoasaphus kowalewski.*
Spectacular trilobites from
Ordovician strata near St.
Petersburg, Russia have entered
the fossil market, some with
spectacular stalked eyes as shown
here. Middle Ordovician, Wolchow
River near St. Petersburg, Russia.
(Value range E).

Fig. 04-69. *Neoasaphus kowalewski*. Another view of this extremely well preserved trilobite with its eyes on stalks. Middle Ordovician, Wolchow River near St. Petersburg, Russia.

Fig. 04-70. *Ampyxina bellutula*. These small eyeless trilobites have entered the fossil market in considerable quantity. Zones of these trilobites, sometimes preserved in clusters, have been found by collectors in the Upper Ordovician, Maquoketa Formation of northeastern Missouri. Some, like the specimens shown here are preserved in a yellow (oxidized) siltstone. Others are found in grey limestone slabs occurring in creek beds. (Value range G, single specimen).

Fig. 04-71. *Ampyxina bellutula*. Specimens of this trilobite preserved in silty, thin limestone beds, when spit by the freeze-thaw cycle, can be beautifully detailed and delicate. Maquoketa Formation, Upper Ordovician. (Value range G).

Fig. 04-72. Large slab of *Ampyxina bellutula*. This trilobite is sometimes found clustered together in groups like this. Such clusters may also be associated with small, carbonaceous objects which were probably films of algae (cyanobacteria?) upon which the trilobites may have fed. Maquoketa Formation, Upper Ordovician, Lincoln County, Missouri. (Value range E).

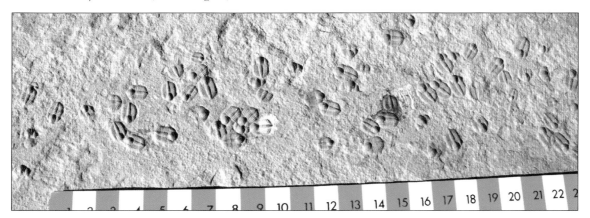

Fig. 04-73. Close-up of slab of *Ampyxina bellutula*. A closer view of one of the clusters of this eyeless or "blind trilobite" from the Upper Ordovician, Maquoketa Formation of Missouri and Illinois. Lincoln County, Mo. (Value range E).

80

Graptolites

Graptolites are problematic fossils which align themselves with some primitive chordates found recently in modern seas. These are typical Ordovician forms from deep sea sediments such as in the Ouachita Mountains of Arkansas.

Fig. 04-76. *Didymograptus* sp. Graptolites, because they were pelagic or floating organisms, have a widespread distribution. They are most commonly found in black, slaty shales, such as these from the Ouachita Mountains of Arkansas. Such shales were probably deposited in the deep waters of the open ocean and not on areas of continental crust as were limestones. The red color of the shale is a consequence of deep weathering and oxidization of the shales. Womble Shale, Crystal Springs, Arkansas. (Value range G).

Fig. 04-74. *Nemagraptus gracilis*, Hall. A particularly graceful graptolite colony which identifies graptolites of the Middle Ordovician. This same graptolite occurs in the Appalachians of Virginia, Tennessee, Georgia and Alabama. The Arkansas sediments in which it occurs were deposited in deeper water than that of the Appalachian occurrences. Black, slaty shales of the Ouachita Mountains are believed to be deep sea sediments which were possibly scraped from an ocean floor by sea floor spreading in the Paleozoic era. Such deep sea sediments like those of the Ouachita Mountains of Arkansas and Oklahoma are some of the oldest deep sea sediments in North America containing fossils. Womble Shale, Middle Ordovician, Crystal Springs, Arkansas. (Value range G).

Fig. 04-75. This is a complex graptolite colony characteristic the Middle Ordovician. Cow Head, Newfoundland.

Fig. 04-77. *Didymograptus* sp. (left) and *Monograptus* sp. (right). Two common graptolites of the Middle Ordovician. Womble Shale. (Value range G).

Vertebrates

These are some of the earliest vertebrates in the fossil record, they are usually pretty scrappy!

Fig. 04-78. *Astraspis desiderata*, Walcott. (Ostracoderm shield fragments) These scrappy bone fragments in red sandstone represent one of the oldest records of vertebrates (Ostracoderms are jawless, bony armor fish). They occur in a red sandstone of Middle Ordovician age uplifted at the front range of the Rocky Mountains near Canon City Colorado. Hardin Sandstone, Middle Ordovician. (Value range F).

Fig. 04-79. *Astraspis desiderata*, Walcott. These fragments of bony armor fish don't look like much, but they "herald" the beginning of the vertebrate fossil record. They were originally found and described by Charles D. Walcott as one of his first discoveries when he was initially sent west with the new U.S. Geological Survey. Hardin Sandstone, Middle Ordovician, Cannon City, Colorado, (Value range H)

Fig. 04-80. Group of sandstone slabs bearing fish plate fragments. Small bone fragments, which look somewhat like these, have been found in the Cambrian of Wyoming, however these are believed to be fragments of arthropods (Aglaspids?) with a similar calcium phosphate composition.

Fig. 04-81. Close up of the yellow slab of previous picture. Hardin Sandstone, Canon City, Colorado. (Value range G).

Bibliography

Finney, Stanley C., and Matthew H. Nitecki, 1979. *Fisherites*, n. gen. *reticulatus* (Owen), 1844. A new name for *Receptaculites oweni* Hall, 1861. *Journal of Paleontology*, Vol 53, No. 3 pp. 750-753.

Palmer, Douglas, and Barrie Rickards, 1991. *Fossils Illustrated-Graptolites, Writing in the Rocks*. Boydell Press, Woodbridge, England.182 pp.

Chapter Five
The Silurian Period

A Wealth of Paleozoic Marine Life

Marine plants and animals make up most of the
life of the Silurian. Land areas of the earth were just
becoming covered with some small plants, but these,
in view of their scarcity, were very localized.
These are all relatively odd Silurian fossils.

Fig. 05-01. These fossils resemble some type of small,
herbaceous aquatic plant, possibly a charophyte. Silurian land
plants are also known but they are small and inconspicuous.
Fossils such as this have been found in many areas of Silurian
rocks and are problematic. Numerous fossils which suggest land
plants occur in the early Paleozoic but determining what many
of them are is difficult. Edgewood Formation, Lower Silurian,
northeast Missouri.

Fig. 05-03. Sponge. These large, bowl-shaped sponges occur in
the Silurian of Quebec where they appear in reefs in what were
probably fairly deep waters. Middle Silurian, Eastern Quebec,
Canada. *Courtesy of A. C. Spreng*, Rolla, Mo.

Fig. 05-02. *Cyclocrinites dactyloides*. These
are Silurian receptaculitids. They were
originally considered to be a type of
crinoid head or calyx. Receptaculitids are
a type of calcareous red algae that are
usually associated with the Middle and
Upper parts of the Ordovician Period.
Hopkington Dolomite, Farmers Creek
Member, Jones Co., Iowa. (Value range G,
single specimen).

Corals

Corals are common Silurian fossils, as corals became abundant marine animals in the Ordovician. Their robust corallites make attractive fossils such as these.

Fig. 05-04. *Halysites catenulatus.* Chain coral, part of a Silurian glacial cobble found in Iowa and probably derived from Silurian strata west of the Canadian Shield in Manitoba transported southward by Pliestocene glaciers. *Halysites* sp. (chain corals) are distinctive, attractive and relatively common Silurian fossils in North America. (Value range G).

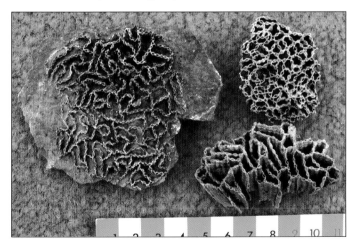

Fig. 05-05. *Halysites gracilus,* Hall. A group of silicified chain corals from Manitoulin Island, Ontario, Canada.

Fig. 05-06. *Halysites gracilis,* Hall. These corallites have been silicified and then were weathered out in relief. Silurian of Manitoulin Island, Ontario, Canada. (Value range G).

Fig. 05-07. *Favosites cf. F. favosus.* A common coral genus of both the Silurian and Devonian. This specimen is silicified. Manitoulin Island, Ontario, Canada. (Value range G).

These are relatively unusual Silurian fossils.

Fig. 05-08. Jellyfish (?) A number of these puzzling fossils from Xuefeng Mountain, Hunan Province came onto the fossil market labeled as jellyfish. These are unlike any described jellyfish known to the author but they do resemble some types of trace fossils. Paleontologist Zang Xinpin suggests a relationship to protomedusa of Walcott, 1899, but many of Walcott's "jellyfish" are now considered to be trace fossils. (Value range F).

Fig. 05-09. *Lecthylus* sp. (Sipunculid or accordion worms) Zones of these fossil "accordion worms" occur in Silurian strata of the Chicago, Illinois, area. Temporary excavations in the windy city have yielded quantities of them from time to time; they were distributed among collectors and museums. The Silurian of the Chicago-Milwaukee region has yielded other horizons of non- or lightly skeletonized invertebrates such as the Brandon Bridge fauna of Wisconsin. (Value range E).

Fig. 05-10. *Nerites* sp. These fossil deep sea trackways were made by some type of deep sea invertebrate which extracted organic matter from the ocean floor. A furrowing strategy is utilized as an energy economizing one by deep sea organisms to maximize efficiency in their extraction of organic material from ocean floor sediments. Blaylock Formation, Lower Silurian, Ouachita Mountains, Oklahoma.

Echinoderms

Echinoderms are some of the more desirable and collectable Silurian fossils.

Fig. 05-11. *Holocystites scutellatus*. These cystoids represent a class of echinoderms which went extinct in the Devonian Period. Cystoids have less regularity in their symmetry than do crinoids. A number of these cystoids have circulated in the fossil collecting community, where they come from a limy shale layer in a quarry in southern Indiana. Osgood Shale, Napoleon, Indiana. (Value range G).

Fig. 05-12. *Sinocystis* sp. These cystoids from China showed up on the fossil market in 2006. They are the interior molds of a cystoid which resembles a hen's egg. The small calcareous plates have dissolved away leaving this internal mold. (Value range F).

Fig. 05-13. *Eucalyptocrinites crassus*. A distinctive calyx of a fairly large crinoid from the Middle Silurian of Southern Indiana. Waldron Shale, Middle Silurian. (Value range F).

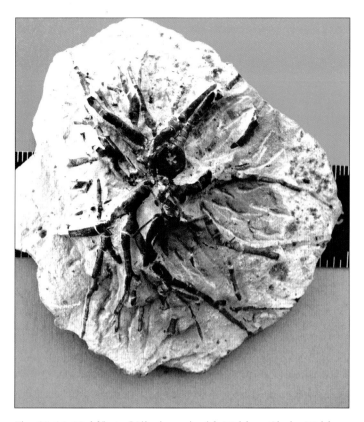

Fig. 05-14. Holdfast of Silurian crinoid. Waldron Shale, Waldron Indiana. These "root-like" holdfasts were probably from *Eucalyptocrinites* with which these holdfasts are associated. (Value range G).

Fig. 05-16. These small, graceful crinoids occur in early Silurian strata of southeastern Missouri. Cape Girardeau Formation, Lower Silurian, Cape Girardeau County, Missouri.

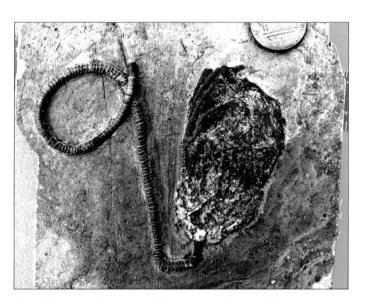

Fig. 05-15. *Ptychocrinus* sp.These graceful Lower Silurian crinoids occur in slabby limestone of southern Illinois and southeastern Missouri. Cape Girardeau Formation, Lower Silurian.

Fig. 05-17. A complete, graceful Silurian crinoid from southeast Missouri. (Value range F).

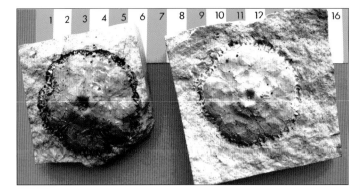

Fig. 05-18. These crinoid calyx's in dolomite are typical of internal casts of crinoids. The stem and arms of the animal is missing. Middle Silurian, Chicago, Illinois. (Value range G, single specimen).

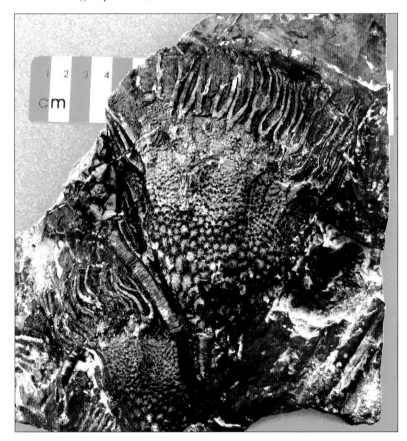

Fig. 05-19. *Scyphocrinus elegans*. These attractive crinoids seem to have been floaters rather than having been attached to the sea floor, as were most crinoids. Associated with *Scyphocrinus* are found partitioned, globular fossils originally thought to be crinoid heads and given the generic name *Camarocrinus* sp. These are now known to have been the floats of *Scyphocrinus*. Upper Silurian, Jissoumour area between Alnif and Tazoulait, Morocco (often given as Erfoud Morocco). Individual specimens from a large colony of these crinoids have been widely distributed through the collector community. Large crinoids, similar to these from Morocco are also known from near the Silurian-Devonian boundary at a number of localities in the northern hemisphere. (Value range F).

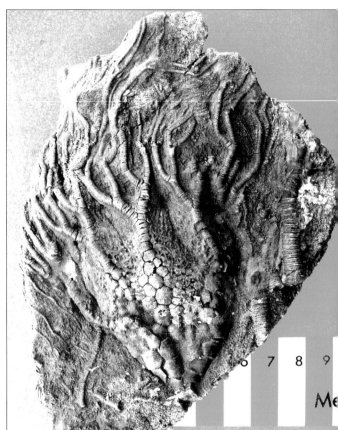

Fig. 05-20. *Scyphocrinus elegans*. A single specimen collected in the 1980s from surface outcrops. Note how the surrounding matrix has more iron oxide stain as a consequence of near surface weathering. Compare this specimen with the previous one which was collected fifteen years later by quarrying into the crinoid rich zone (Value range F).

Fig. 05-21. *Scyphocrinus elegans*. A single individual specimen of *Scyphocrinus* sp. from Morocco. These are some of the Moroccan Paleozoic fossils which have become available at a reasonable price. Many of the Paleozoic fossils found in Morocco have also been found in similar age rocks of Europe, however the Moroccan specimens are of better quality, are more readily available, and are reasonably priced. Crinoids like this one from other parts of the world would bring a much higher price. (Value range D).

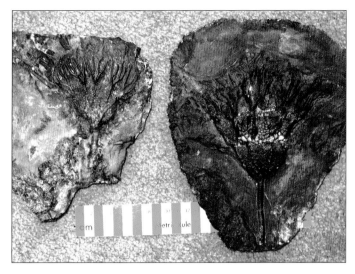

Fig. 05-22. Two specimens of *Scyphocrinus elegans* from Pridolian strata of Morocco. The large amount of Paleozoic strata that occurs in the Atlas Mountains accompanied by the opening of quarries (for fossils) in the fossil bearing layers, has allowed large numbers of these excellent specimens to be collected and placed on the fossil market at competitive prices. This occurrence of *Scyphocrinus* appears to be widespread. Specimens from Morocco are often labeled as being Devonian in age, however they can just as well be placed at the end of the Silurian Period as has been done here. *Scyphocrinus* occurs worldwide at or near the Silurian-Devonian boundary. (Value range E).

Fig. 05-24. *Scyphocrinus missourensis.* Two calyce's of this large crinoid which lacks the arms and stem. Bailey Formation, Upper Silurian. North of Cape Girardeau, Missouri. (Value range E).

Fig. 05-23. *Scyphocrinus missourensis*, Springer. Part of a colony of this crinoid which occurs along the Mississippi River north of Cape Girardeau Missouri. A large slab of this crinoid colony resides in the U.S. National Museum, Washington D. C. where it was collected in 1911 by associates of Frank Springer, a pioneer crinoid worker. Associated with the U S National Museum slab are the spherical partitioned structures given the name of "*Camarocrinus.*" These have been found to have been the hollow floats to which *Scyphocrinus* was attached. Bailey Formation, Upper Silurian-Lower Devonian. North of Cape Girardeau, Missouri. (Value range D).

Fig. 05-25. *Camarocrinus* sp. These floats of *Scyphocrinus* are also referred to as loboliths. They originally were given the generic name *Camarocrinus* sp. but were later recognized to be floats to which the crinoid *Scyphocrinus* was attached. Large numbers of these have come from strata of the Arbuckle Mountains of Oklahoma, as did these. Note the crinoid stem protruding from a collar on the specimen at the left. Harrigan Formation, Arbuckle Mountains, Oklahoma. (Value range G, single specimen).

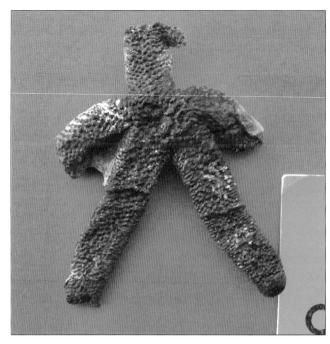

Fig. 05-27. Starfish, Cape Girardeau Formation, southeast Missouri.

Fig. 05-26. Starfish, *Palaeaster* sp. A Paleozoic starfish from Lower Silurian strata near the Ordovician-Silurian boundary. Girardeau Formation. (Value range F).

Fig. 05-28. *Urosoma* sp. Two sandstone slabs bearing impressions of specimens of this starfish from the lower Silurian of eastern Australia. These came into the fossilphile community about 1985. Paleozoic ophuroids (brittle stars) and asteroids (starfish) are relatively rare fossils so that these multiple specimen slabs are unusual and desirable (Value range E).

Fig. 05-29. *Urosoma* sp. Single slab of sandstone with starfish impressions from Silurian sandstone of eastern Australia.

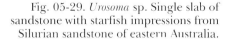

Mollusks

Mollusks such as these cephalopods are desirable Silurian fossils.

Fig. 05-30. *Orthoceras vertebrale.* This ribbed, straight nautaloid cephalopod is characteristic of both the Middle and Late Ordovician as well as of the Silurian. The species name suggests a string of vertebra which the cephalopod resembles. Joliet Dolomite, Grafton, Illinois. (Value range G).

Fig. 05-31. *Crytoceras* sp. This coiled nautaloid is a somewhat abundant Silurian Cephalopod. Marine Silurian and Devonian strata can yield a variety of straight, partially coiled (like this), and coiled nautaloid cephalopods. Middle Silurian dolomite beds of the Joliet Formation near Grafton, Illinois, as well as the type locality of the formation near Joliet, Illinois, has produced many nice fossils over the years which have found their way into collections. (Value range F).

Arthropods

Arthropods of the Silurian include ostracods, phyllocarids, and, of course, trilobites of which the calymenids are particularly prominent. Eurypterids are also highly desirable Silurian fossils.

Fig. 05-32. *Herrmonnia* sp. Ostracodes, also known as clam-shrimp, are arthropods which still live today, although they usually are quite small. Ostracodes are important and abundant microfossils; the specimens shown here are giants by comparison. From slaty shales, Newbury Formation, Rowley, Mass. (Value range G).

Fig. 05-33. These two slabs of even larger ostracode impressions (compared with the above image) come from slaty shale of the Newbury Formation, Rowley, Mass.

90

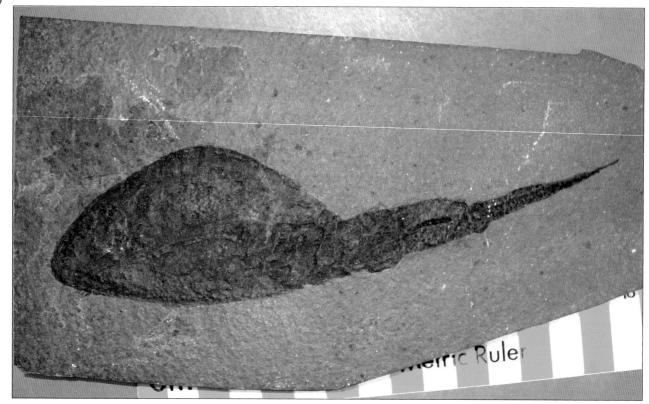

Fig. 05-34. *Ceratiocaris papilo* (a Phyllocarid). The phyllocarids are a group of extinct shrimp-like arthropods. Paleozoic arthropods other than trilobites and ostracodes are generally uncommon fossils. Priesthill Group, Silurian, Dunside. Lesmahagow, Lanarkshire, Scotland. (Value range D).

Fig. 05-35. *Gravicalymene (Flexicalymene)* cf. *tazarinensis*. As with many widely distributed trilobites from Morocco, the nomenclature on them is confusing. They are calymenid trilobites originally given the name of *Calymene tristani* from European specimens in the 19th century. As the Moroccan Paleozoic rocks and their fossils are closely related to those of Europe, this name is reasonable and somewhat appropriate. However the genus *Calymene* has been split into a number of genera such as *Gravicalymene* and *Flexicalymene*, both of which reliable sources have identified as the currently accepted genus of this trilobite. The genus *Stenocalymene* also comes up in relationship to the Moroccan specimens, however this genus appears to be exclusively related to the North American craton. Middle(?) Silurian, Eufrod, Morocco. (Value range H).

Fig. 05-36. These calymenid trilobites are found in great numbers where they have weathered from shale beds in the intermountain valleys of the Atlas Mountains of Morocco. They are often collected by children who scour the dry shale slopes, who, at an early age, are introduced to the world of fossil collecting. Trilobites of the genus *Calymene* were compact and tough, they did not separate into head (cephalon), thoraxial segments and tail (pygidium) very easily as most other trilobites did upon molting. Specimens of these trilobites are variously labeled as Ordovician as well as Silurian. Most trilobites of the genus *Calymene* are from the Silurian. (Value range H).

Fig. 05-37. Moroccan calymene specimens occur in concretions where part and counterpart occur as shown here. This is also the most common occurrence of Moroccan calymenids, as single "bugs." As a consequence of the large number of specimens exported from Morocco, the market has become saturated with them, so that the price of these trilobites is quite low. After initial collecting (which in the case of Moroccan calymenids has extended over three decades), the law of diminishing returns takes hold and the fossils become harder to find; this, then, usually brings up the value of the item. Moroccan *calymene* trilobites occur over such a large area in the High Atlas that they keep coming onto the fossil market. When the "easy pickin's" end they should increase in value. (Value range as of writing H).

Fig. 05-38. Group of Moroccan calymene trilobites. Probably no trilobites have been more widely distributed than these. Found weathered from shale beds in populated valleys between mountains of the High Atlas of Morocco near the city of Eufrod, these trilobites are collected by locals, often children, from the silty desert regolith. The geologic age of these trilobites is often listed as late Ordovician, but a Silurian age is most appropriate as the genus *Calymene* achieves its maximum abundance during this time. Almost identical specimens to these are found in Silurian silty sandstones of the southern Appalachian Mountains in the states of Georgia and Alabama. This underlines the fact that eastern North America, Europe and North Africa were once part of the same continental mass prior to the break up of Laurentia. It was continental drift that separated these areas from each other with the consequent formation of the Atlantic Ocean and the Mediterranean Sea after the Paleozoic Era.

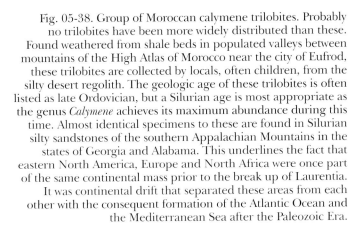

Fig. 05-39. This specimen of *Calymene* sp. is virtually identical to the Calymenids from Morocco. This specimen, has been foreshortened by being preserved in slate where pressure from mountain building has squeezed and changed the original shale to slate, a low-grade metamorphic rock. Southern France. (Value range G).

Fig. 05-40. *Calymene (Stenocalymene) celebra*. Large numbers of this species of *Calymene* have come from Middle Silurian Joliet Dolomite of Illinois as well from similarly aged dolomite elsewhere. These trilobites were locally known as "rock dogs" by quarry workers and locals. Joliet Dolomite, Grafton, Illinois. (Value range H, single specimen).

Fig. 05-41. *Calymene (Stenocalymene) celebra*. Left: Joliet, Illinois; Right: Grafton, Illinois. The trilobites of the genus *Calymene* have been split into a number of genera such as *Stenocalymene* on the basis of variations in eye sutures. What look like eyes on calymenid trilobites are not; they are what are known as papibral lobes. Calymenid eyes were quite small and inconspicuous. This common calymenid trilobite found in Silurian Dolomite of Illinois has been placed in the genus *Stenocalymene* by paleontological splitters. Species are defined in biology as a member of a population of organisms capable of breeding and reproducing. Such a definition obviously cannot be applied to fossils so that what defines a species (and genus) is based upon morphology (shape). Some paleontologists (splitters) consider small shape variations such as minor variations in the eye sutures on calymenes as representing different species. Such splitters split the genus *Calymene* into new genera based upon such criteria. Other paleontologists (lumpers), don't go along with this reasoning. They say such small variations represent normal individual variations among a population of trilobites. Such interpretations and arguments fill quite a lot of scientific literature and produce confusion among collectors who want the "correct" scientific name. Sorry folks, but it's not so "black and white."

Fig. 05-43. *Eoharpes* sp. A distinctive trilobite order characteristic of a trilobite with a wide brim on the cephalon. Edgewood Dolomite, Lower Silurian, northeast Missouri. (Value range F).

Fig. 05-44. *Tetrataspis* sp. A pygidium ("tail") of this distinctive spiny Silurian trilobite. Trilobites with similar pygidia have come from the Atlas Mountains of Morocco and are known under the genus *Crotocephalus*. Bainbridge Formation, southeast Missouri. (Value range G).

Fig. 05-42. *Dicalymene* sp. The presence of original exoskeleton material is what makes these calymenids so desirable. A relatively large number of these trilobites are in the collecting community, often having come through Geological Enterprises of Ardmore, Oklahoma, where, in the 1960s, they worked pits for them in the Arbuckle Mountains. These calymene trilobites are found in siltstone and have original hard part preservation unlike the calymenids of Morocco and the U.S. midwest. Henryhouse Formation, Ada, Oklahoma, (Value range F).

Fig. 05-45. *Dalmanites limulurus* (part and counterpart). These trilobites are found in a Silurian shale bed in some quantity in upstate New York. Rochester Shale, Orleans County, New York. (Value range F, both halves).

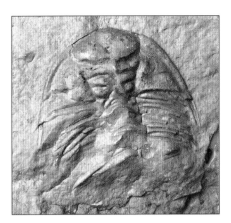

Fig. 05-46. *Dalmanites* sp. The trilobite genus *Dalmanites* is a widespread Silurian genus. These come from Lowermost Silurian strata (Edgewood Formation) of northeastern Missouri. A number of nice cranidia have entered the fossilphile community, but complete specimens are very rare in these dolomitic limestones.

Fig. 05-47. *Coronocepalus changni*. This Silurian trilobite from China represents a genus not found in the western hemisphere, except in the Kalmath Mountains of northern California, where specimens similar to this have been found in marble. Xiushang Formation, Shantung Province, China. (Value range G).

Fig. 05-48. *Cruzania* sp. A peculiar form of the trace fossil *Cruzania* (specimens often labeled as *Rusophycus*). Trace fossils such as these were probably made by trilobitomorphs (soft bodied trilobite-like arthropods). A number of these showed up at the Tucson show in 2001. *Rusophycus* is much more elongate than this trace fossil. Hanyan, Hubei, China. (Value range G).

Fig. 05-49. *Eurypteris lacustris*, Harland. Eurypterids are usually rare and infrequently found fossils. Upper Silurian limestones of upstate New York and adjacent parts of eastern Ontario are an exception. Outcrops of the eurypterid bearing Bertie Formation have yielded many eurypterid specimens for over a century, however in the 1980s, the opening of pits to mine these fossils made them much more accessible and available along with associated fossils, some of which were never known before. This is a case where commercial fossil collecting of a well known fossil benefited paleontology by unearthing previously unknown organisms.

94

Graptolites

Graptolites, an extinct group of hemichordates, are found in deep sea, Silurian slaty shales.

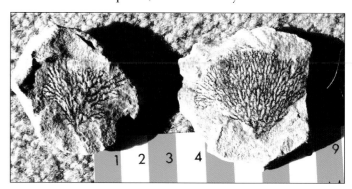

Fig. 05-52. *Dictyonema ratiforme*. Nice slabs of these denderoid graptolites come from the Silurian Gosport channel lens in the Lockport Formation in western New York state. (Value range G).

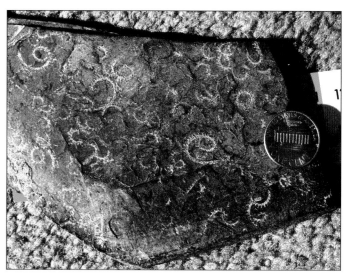

Fig. 05-53. *Spirograptus* sp. These coiled graptolites are typical for the Silurian. Graptolites like this lived at the surface of the ocean and are found in deep sea sediments where colonies of these organisms floated on the surface. Graptolites are now believed to be a group of extinct hemicordates. They are generally preserved as shiny carbon films in slaty shale. These black, slaty shales usually are representative of deep sea sediments, often in trenches, where few other organisms are found as fossils. Graptolite colonies were pelagic or floating organisms which were preserved by the anoxic nature of deep sea sediments.

Fig. 05-50. *Eurypterus lacustris*, Harland. A group of four eurypterid specimens from the Bertie Formation of northern New York. Passage Gulf near Cedarville, Herkimer County, New York (Value range D).

Fig. 05-51. *Eurypterus* sp. Eurypterids can be locally frequent in these grey, fine grained limestones known as water lime. The fine grain of the sediment preserves the delicate, chitonous eurypterid exoskeleton quite superbly. Other, more normal limestones, rarely preserve eurypterids or if they do they are very faint. Bertie Formation, Upper Silurian. (Value range C).

Fig. 05-54. *Monograptus turrculatus*. Silurian graptolites can be coiled such as this. This type of coiling is absent in earlier forms. Lower Silurian, Planer, Germany.

These are the head shields of jawless, bony armor
fish, some of the earliest known vertebrates

Fig. 05-55. These ostracoderm (or Osteostraci) dorsal shields represent some of the earliest known fishes. Ostracoderms were jawless fishes which lived in areas of brackish water. Left: head shield with the elongate pits are believed to be related to the "electric organs" similar to those found in modern eels. Right: dorsal shield with bone cells. Middle Silurian, Blain, Pennsylvania. (Value range F).

Fig. 05-56. "Ostracoderm" dorsal shield with "electric organ" impression.

Fig. 05-57. Ostracoderm with dorsal shield with part of the shield exfoliated showing bone cells. Bone in early vertebrates was restricted to an outside bony armor, the internal notochord was not mineralized with calcium phosphate, as would be the case with later vertebrates. Middle Silurian, Blain, Pennsylvania. (Value range F).

Bibliography

Hess, Hans 1999. "Scyphocrinitids from the Silurian-Devonian Boundary of Morocco" in *Fossil Crinoids*. Cambridge University Press, Cambridge, England.

Chapter Six

The Devonian Period

The Age of Fishes

The Devonian Period found major parts of the earth's land areas covered for the first time with vegetation. Marine invertebrates such as trilobites, corals, cephalopods and crinoids lived in the seas while sharks and shark-like fishes appeared for the first time. A variety of fishes lived in brackish waters, including lobe-finned fish (crossopterygians), lung fish, and placoderms (bony armor fish with jaws). The appearance of this diversity of fish types for the first time has led the Devonian to be called the "age of fishes."

Land Plants

These are all examples of Devonian land plants.

Fig. 06-01. Psilophytes. These are primitive, leafless plants from one of the localities in eastern Quebec. First described by J. W. Dawson, they are some of the earth's first undoubted land vegetation. Such primitive, leafless plants are known as psilophytes. Lower Devonian, Cape aux Os, Gaspe Peninsula, Quebec.

Fig. 06-03. Psilophytes(?) Shale slabs covered with compressions of these leafless plants (presumed psilophytes) can be locally common south of St. Louis, Mo., Grassy Creek Formation, Upper Devonian.

Fig. 06-04. *Archaeopteris* sp. A "pre-fern," *Archaeopteris* is the fern-like vegetation of an early tree of the Devonian Period. It is found in rift valley sediments in the eastern U.S. and in the Canadian Maritimes, where it can be locally abundant. *Archaeopteris* represents one of the oldest land plants that has distinctive and attractive foliage. Frasnian, Middle Devonian, Youngsville, New York. (Value range F).

Fig. 06 02. Psilophyte (?) These leafless plants are believed to be psilophytes, although the occurrence has not been extensively investigated. They occur in late Devonian strata along the northeast edge of the Ozark Uplift. Grassy Creek Formation, Upper Devonian, Hillsboro, Jefferson Co., Missouri. (Value range G).

Fig. 06-05. *Archaeopteris* sp. *Archaeopteris* is associated with the earliest known fossil logs of the genus *Callixylon*. It is believed to be the foliage of what some paleobotanists hypothesized may have been a primitive type of seed plant. Lower Devonian, Youngsville, New York.

Fig. 06-08. Part and counterpart of Late Devonian fern. Escuminac Formation, Escuminac, Quebec.

Fig. 06-07. "Ferns." A number of ferns or fern-like vegetation different from *Archaeopteris* appears in the late Devonian. These compressions are of fern-like foliage from Late Devonian rift deposits at Escuminac, Quebec.

Fig. 06-06. *Archaeopteris*. These early land plants are generally preserved in a type of dirty sandstone known as greywacke. Greywacke is often the type of sediment which filled what is known as a rift valley (the early stages of a rift zone). These specimens of *Archaeopteris* come from greywacke in eastern Quebec, Canada. Escuminac Formation, Upper Devonian, Escuminac, Quebec.

Fig. 06-09. *Callixylon* sp. A slice through a petrified log of what may have been the trunk of the tree which bore *Archaeopteris* foliage. (Value range F).

98

Sponges

Sponges (parazoans) are the phylum represented by figures 10 through 13.

Fig. 06-10. Stromatoporoids. Silicified stromatoporoids from glacial deposits in northern Missouri. These specimens probably came from Devonian strata in Iowa where they were brought down by glaciers during the ice age. Stromatoporoids are a type of sponge which at one time was confused with stromatolites, which are structures produced by blue-green algae. (Value range F, single specimen).

Fig. 06-11. *Armstrongia oryx* (*Titusvillia*). This is the impression in Devonian greywacke of a sponge colony that lived in seaways covering much of what would later become the Appalachian Mountains. Such greywacke was deposited in shallow seas; the greywacke in which *Archaeopteris* and early fish were preserved was deposited in fresh or brackish water and lack marine fossils such as this. (Value range F).

Fig. 05-12. *Astraeospongia* sp. These distinctive Devonian sponges are characterized by being composed of "star-like" spicules. Middle Devonian, western Tennessee. (Value range G).

Fig. 06-13. *Hindia sphaeroiidalis*. A common sponge in Lower Devonian marine strata of Western Tennessee. Beech River Formation, Decatur Co., Tennessee.

Corals

Corals can be frequent Devonian fossils, especially on the craton.

Fig. 06-14. *Acrophyllum* sp. A typical Devonian coral that lived in the seas covering the craton and harboring numerous coral reefs. This specimen came from Morocco but similar fossil corals are found in Devonian marine strata worldwide. Middle Devonian, Atlas Mountains, Morocco. (Value range G).

2</

99

Fig.06-15. *Pachyphyllum nevadens*. A colonial tetracoral from marine Devonian rocks of Arizona. Pine County, Arizona (Value range G).

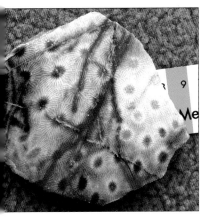

Fig. 06-16. *Phillipsastraea hennahi*. A section through a colonial tetracoral from marine strata of the type area (area exposing the rocks for which the period was named) of the Devonian Period. Middle Devonian, Devon, England. (Value range F).

Fig. 06-19. *Hadrophyllum orbignyi*. Appropriately called "button corals." Middle Devonian, Speed, Indiana. (Value range G).

Fig. 06-17. *Syringopora* sp. A coral genus common in the Devonian and Mississippian periods. Middle Devonian, Brooks Range, northern Alaska. (Value range H).

Cnidaria

Figures 20 through 23 are problematic fossils which some paleontologists "shoe horn" into the phylum cnidaria.

Fig. 06-20. *Plumalina plumaria*, Hall. A problematic fossil originally considered to be a plant. *Plumalina* is now considered as a type of soft coral or sea pen. Upper Devonian, western New York state. Value range G).

Fig. 06-18. *Zaphrentis* sp. A cast of the interior showing septa of this solitary tetracoral. Being a tetracoral, the number of septa shown here would be in a multiple of four. The tetracorals went extinct at the end of the Paleozoic Era; post Paleozoic corals have their septa in multiples of six (hexacorals). Devonian outlier, Rolla, Missouri. (Value range G).

Fig. 06-21. *Plumalina* sp. Preserved in a dirty sandstone (greywacke).

Fig. 06-22. *Conularia africana*. A problematic Paleozoic organism the has been placed in the class Cnidaria, the mollusca, and, by some paleontologists, even into an extinct animal phylum. Icla Shale, Belen, Bolivia. (Value range F).

Brachiopods

Brachiopods can be well represented in Devonian marine strata.

Fig. 06-24. *Schizophoria* sp. This genus of brachiopod is particularly characteristic of the Devonian Period. Its name comes from Schizo=dual, in reference to the very different shell morphology of the dorsal (top) valve compared with that of the ventral (bottom) valve. Devonian outlier, Owensville, Missouri. (Value range F, single specimen).

Fig. 06-25. Typical Devonian brachiopods. A selection of frequently found brachiopods characteristic of Devonian marine strata.

Fig. 06-23. *Conularia africana*. Additional specimens of this problematic Paleozoic fossil. (Value range G, single specimens).

Tentaculites are another problematic Devonian fossil.

Fig. 06-26. *Tentaculites* sp. *Tentaculitids* are problematic fossils. They have been considered as mollusks, as a member of some worm phylum, as well as other designations, They are almost exclusively Devonian in age. Czortkov, Ukraine. (Value range F).

Fig. 06-27. *Tentaculites* sp. A common fossil in Devonian marine strata of many regions. Cedar Valley Limestone, Coralville, Iowa.

Echinoderms

Echinoderms are always desirable fossils, particularly when they are complete.

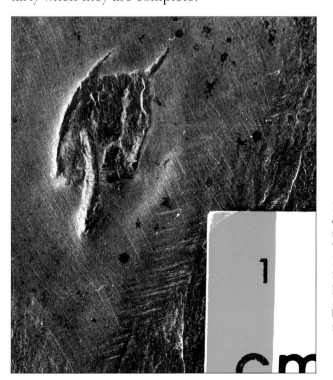

Fig. 06-29. Two cystoids from the Bundenbach slate of Germany. Left: *Rhenocystis* sp.; Right: *Regulaecystis pleurocystoides.* Lower Devonian.

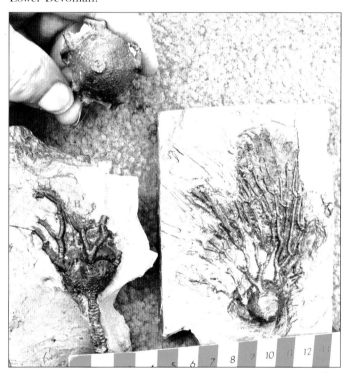

Fig. 06-30. *Arthroacantha carpenteri*. A group of average specimens of this crinoid, often found pyritized in grey shale. Middle Devonian, Sylvania, Ohio. (Value range F, single specimen).

Fig. 06-28. *Rhenocystis* sp. This Devonian cystoid is one of the last examples of this extinct Echinoderm Class. The specimen is in the famous Bundenbach slate, a Lower Devonian paleontological "window" in the Rheinish Mountains (Black Forest) of Germany. Bundenbach or Hunsruck fossils are pyritized (apparently before metamorphism), and this pyrite replacement can include the preservation of soft tissues such as trilobite legs and digestive system. Bundenbach fossils come from a series of slate quarries where it has been quarried for over two hundred years. They are quite desirable fossils. (Value range F).

102

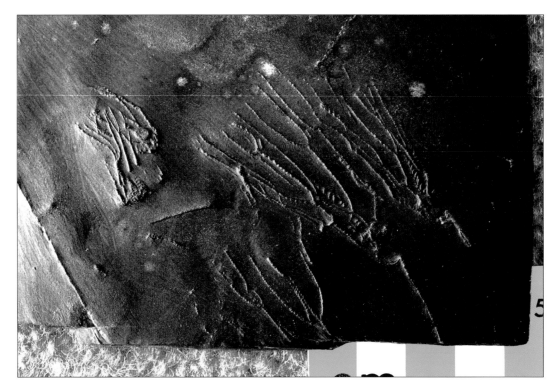

Fig. 06-31. *Parisangulocrinus zeaeformis*. This group of delicate crinoids in slate from Hunsruck or Bundenbach Germany are indicative of a deep water environment under which the (origin shale) accumulated. These fossils are actually found in a low grade metamorphic rock, however the fossils had been replaced with pyrite before metamorphism took place so that they retain amazing detail. (Value range D).

Fig. 06-32. These Bundenbach crinoids probably lived in relatively deep water, an environment which today harbors a number of delicate living crinoids as well as numerous starfish.

Fig. 06-33. This is a crinoid found in the Devonian of both Europe and Morocco. It resembles a green pepper, particularly when not flattened as this specimen is. Lower Devonian, Morocco. (Value range D).

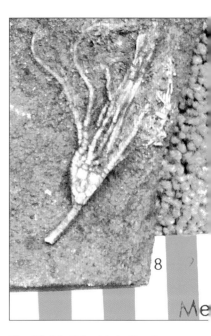

Fig. 06-34. Crinoids in Sandstone. Fossils in sandstone are usually preserved as internal molds. Here these crinoids retain their original mineral composition. Bushberg Sandstone, Upper Devonian, Jefferson County, Missouri.

Fig. 06-35. Left, top and bottom: *Furcaster gracilus*. Bottom right: *Encrinaster roemeri* (webbed starfish). Upper right: *Furcaster paleozoicus*. These Bundenbach starfish show the fine preservation for which the Bundenbach slates are famous. (Value range D, single specimen).

Fig. 06-36. *Furcaster paleozoicus*. A quartz vein cuts through this Bundenbach starfish slab. Starfish are one of the more common fossils from this unique fossil locality. Such starfish are believed to have lived in relatively deep waters (as many starfish do today). This is the same specimen as in the upper right of the previous image. (Value range D).

Mollusks

Mollusks (gastropods and cephalopods) are the invertebrate fossils represented here.

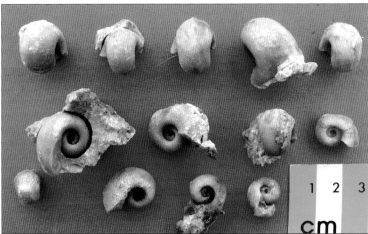

Fig. 06-37. Bellerophonts. This group of small bellerophont gastropods are preserved in quartzite. Bellerophonts are planispiral gastropods that are characteristic of the Paleozoic Era. The illustrated specimens came from an isolated occurrence (outlier) of Devonian rock in the Ozark Uplift. Ozark outliers contain a greater that normal number of fossil mollusks as the waters in which the original sediments were deposited were of greater than normal salinity. Mollusks can tolerate such saline conditions better than most other marine invertebrates. Middle (Lower according to J. Bridge) Devonian outlier, Rolla, Missouri. (Value range G, single specimen for similar material).

Fig. 06-38. These gastropods are preserved as internal molds (steinkerns) in quartz sandstone. From another Ozark Devonian outlier (see above caption). Owensville, Missouri.

Fig. 06-39. These are the "lids" or opercula of gastropods. They were originally composed of a leathery material which left these impressions in sand. Such impressions are rarely found in limestone. Middle Devonian outlier, Owensville, Missouri.

Fig. 06-40. "Orthoceras." These are some of the most common and ubiquitous fossils coming from Morocco. They are straight nautaloid cephalopods of a type common throughout much of the Paleozoic Era. (Value range H).

Fig. 06-42. "Orthoceras." The black limestone which contains these straight cephalopods from Morocco is used as interior "marble" for interior construction. These are two marble scraps containing straight cephalopods. (Value range H).

Fig. 06-41. "Orthoceras" slab. These straight cephalopods are aligned parallel with each other in a manner often seen with other straight cephalopod occurrences. They are thought to have been aligned in this manner by water currents. Large slabs covered with these cephalopods are obtained in quantities from the Middle Devonian of the Atlas Mountains of Morocco. (Value range F).

Fig. 05-43. Coiled ammonoid (Goniatite). Nautaloid cephalopods have simple, watch-glass shaped partitions of the shell. Many of the coiled cephalopods from the Devonian of Morocco have their shell partitions crenulated or wavy. Those cephalopods which have such a configuration are known as goniatites, which is an early type of ammonoid cephalopod. The goniatites evolved into the ammonites, a major group of animals of the Mesozoic Era. These Moroccan and other Devonian cephalopods with their goniatitic sutures are the first of their kind. (Value range F).

Opposite
Fig. 06-45. *Beloceras sagittrium*. This is a coiled ammonoid cephalopod. Such cephalopods were the beginnings of the ammonites, a dominant group of invertebrates of the Mesozoic Era. Lower Devonian, Atlas Mountains, Morocco.

Trilobites

Trilobites are diverse and sometimes ornate in the Devonian.

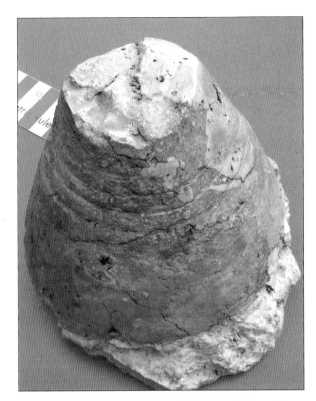

Fig. 06-44. Funnel or phonograph horn cephalopod. This cone-shaped cephalopod from Devonian rocks of Central Manitoba is part of an "Arctic" Paleozoic fauna. The northern part of North America yields Paleozoic faunas which are often different from those found farther south. The shape of this cephalopod resembles that of the horn of an Edison tinfoil phonograph. Cephalopods with such a shell shape are absent from Devonian faunas of the southern part of the continent. Middle Devonian, The Pas, Manitoba, Canada.

Fig. 06-46. *Proetus* cf. *P. searighti*. The genus *Proetus* is a ubiquitous Devonian trilobite in marine Devonian rocks. This specimen is an impression in quartzite from an anomalous Devonian locality near Rolla, Missouri. In the 1920s, a small, fossiliferous outcrop of quartzite northwest of Rolla was discovered. The fossils present were definitely of Devonian in age but Devonian rocks were unknown within 100 miles of Rolla at that time. J. Bridge, of the Missouri School of Mines in Rolla, considered these fossils to be Lower Devonian in age. He believed that the Rolla quartzite represented a remnant of a once existing extension of Devonian strata from over 100 miles to the east. Later in the 1950s, other Middle Devonian outliers (or remnants) were found north of Rolla which seemed to connect with Middle Devonian strata found north of the Missouri River. Another odd association of these "out of place" fossils regards the species name of this genus, *Proetus*. It was named after Walter Searight, a geology student at the University of Iowa in the mid 1920s. Searight found a Proetid trilobite in the Iowa City area (Middle Devonian, Cedar Valley Limestone) which he gave to one of his professors who then named it after him. Searight later became a geologist for the Missouri Geological Survey in Rolla, Missouri where this trilobite was eventually found.

Fig. 06-47. *Dipleura dekayi* Green. This is a large, spectacular trilobite found in black, fine grained sandstone (greywacke) of central New York state, particularly around Ithaca. (Value range D).

Fig. 06-48. *Phacops africanus* (*Geesops* also *Drotops*). These large phacopid trilobites come from the Atlas Mountains of Morocco in some abundance (particularly desirable for such a large, complete trilobite). The specimens are in black limestone with the original carapace preserved, although this sometimes is not easy to see. All are highly spinose (spiny) and specimens showing these spines have been given the species name of *P. speculator*. Probably no other Moroccan trilobite has seen such a confusion of names than has this one, It is also been called *Drotops*, *Geesops*, and *Phacops africanes*.

The author has encountered specimens of this trilobite which were over-prepared or reconstructed as well as ones which were cleverly bonded to chunks of the black limestone in which the trilobites are found. With these trilobites, probably more than any other Moroccan fossils, the word is "Caveat Emptor." Lower Devonian, Hamar Laghdad Formation, Alnif, Morocco. (Value range F, actual specimen).

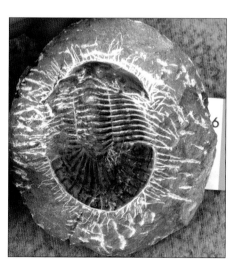

Fig. 06-50. *Phacops africanus* (*Geesops*). Two enrolled (one only partially) specimens of this spectacular trilobite from Morocco. The specimen on the right is a cast or "fake." It is a well made, resin cast skillfully bonded to the black limestone in which the trilobites are found. If one examines the specimen closely with a hand lens or binocular microscope, small bubbles on the surface of the cast can be seen. (Value range, as cast, G).

Fig. 06-51. *Platysacutellum* sp. Hamar Laghdad Formation, District of Alnif, Morocco.

Fig. 06-52. *Crotalocephalina* cf. *globifrom*, Hawle and Corda. These trilobites are also labeled under the genus *Crotalocephalus*. It is one of the more frequently seen of the spectacular Moroccan Devonian trilobites. Hamar Laghdad Formation, Lower Devonian, Erfrod, Morocco. (Value range F).

Fig. 06-53. *Crotalocephalina* (*Crotalocephalus*) sp. Specimens of this genus are often partially enrolled and can appear foreshortened, others have a "broken back." Hamar Laghdad Formation, Lower Devonian, Erfrod, Morocco.

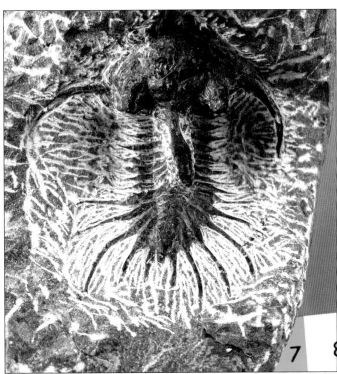

Fig. 06-55. *Philonix* sp. This is one of the very spiny trilobites from the Devonian of Morocco. Trilobites of the genus *Philonix* are also known from Europe. The rocks of the Atlas Mountains are essentially an extension of the same Devonian rocks which occur in Europe, and the same Paleozoic seaways which covered North Africa also covered much of Europe. The formation of the Mediterranean sea with its separation of Europe from Africa came in the Cenozoic Era, much later than the Paleozoic. Hamar Laghdad Formation, District of Alnif, Atlas Mountains, Morocco. (Value range F).

Fig. 06-54. *Paralejurus dormitzeri*, Barrande. One of the other (see above) unusual trilobites frequently seen from Morocco. This genus is also known from Spain and France. Its author, M. Barrande, was a 19th century Bohemian paleontologist who documented a large number of European Paleozoic fossils. Many of the trilobites found in Morocco are also found in Europe and were first described from rocks of southern and central Europe. Hamar Laghdad Formation, Lower Devonian, Alnif, Morocco. (Value range F).

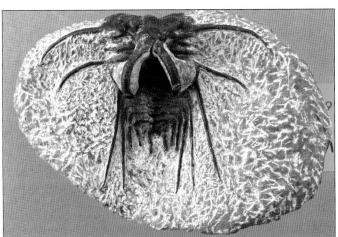

Fig. 06-56. *Dicranurus monstrosus*, Barrande. This is another of the excessively spiny trilobites of the Devonian. The genus *Dicranurus* is found in the Lower Devonian limestones of central Europe in France and Germany, particularly in the Eifel region of Germany and in Bohemia. Many of the Moroccan specimens of *Dicranurus* have been hastily prepared, however specimens of this genus when cleaned with an air abrasive machine can be super-spectacular. They are also quite a bit more expensive to compensate for the hours required in such cleaning. Megrane Formation, Lower Devonian, Atlas Mountains, Morocco.

Vertebrates

Vertebrates diversify and become quite successful in the Devonian.

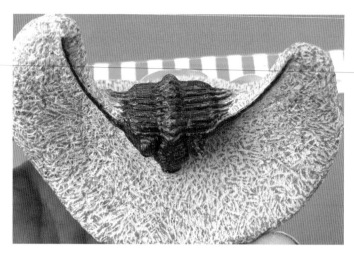

Fig. 06-57. *Psycopyge elegans*, Termier. Another spinose trilobite from the Lower Devonian of the Atlas Mountains. Hamar Laghdad Formation, District of Alnif, Morocco. (Value range E).

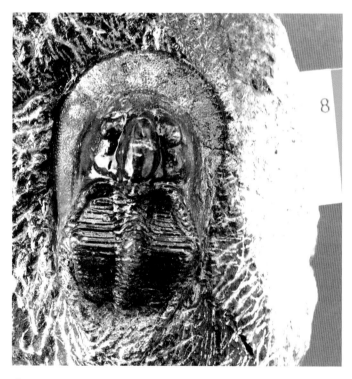

Fig. 06-58. *Harpes* sp. Members of this trilobite family (the Harpidea), are characterized by a wide brim on the cephalon. Merg tgel Formation, Lower Devonian, District of Erfrod, Morocco.

Fig. 06-59. A reproduction of a bizarre comurid trilobite. This undescribed (as of 2006) trilobite is peculiar in its possession of a long, trident-like projection off of the cephalon. This particular specimen was purchased at the 2007 Tucson show for a relatively small sum. It is a reproduction...that is a fake! Rumors have circulated that all or most of the Moroccan trilobites are such reproductions. This is not true! If one examines the actual trilobite (not the rock on which the trilobite cast has been carefully bonded), with a hand lens or binocular microscope, small bubbles in the resin cast can be seen. This is a really nice cast of a very rare trilobite and was worth the $30.00 paid to a Moroccan fossil dealer for it. If Moroccan dealers would indicate which specimens are casts, the credibility of these trilobites (and their desirability) would be maintained; but then again they might all become higher priced. The resin cast is neatly bonded to a chunk of the black limestone in which the trilobites are found.

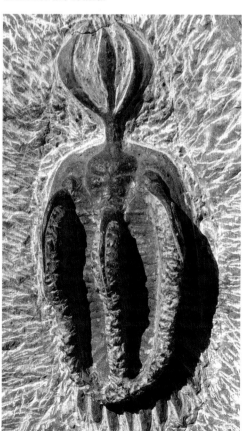

Fig. 06-60. Close-up view of the previously illustrated trilobite reproduction.

Fig. 06-61. *Dicranurus monstrosus*. A resin reproduction (fake) of this interesting Lower Devonian trilobite. At the 2007 Tucson show, one Moroccan dealer had a table of these trilobites which were all identical, but the price was right, $25.00. If one were to sell fake trilobites, don't put out multiples that are all the same...put out one at a time! The center of the specimen (that part made of resin) with the trilobite in the center, is of a lighter shade of buff. This resin center was then set in an oblong chunk of Devonian limestone. (Value range F, as a good reproduction).

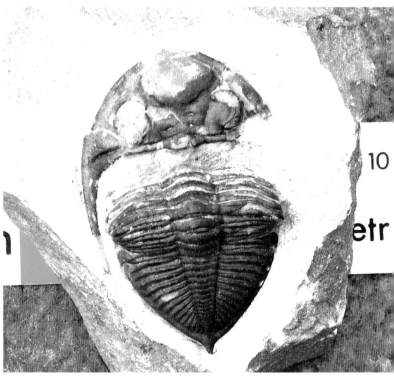

Fig. 06-63. *Odontochile spiniferum*, Barrande. The genus *Odontochile* is a widespread Devonian genus. In North America the genus *Dalmanites* applies to very similar trilobites. Hamar Laghdad Formation, Lower Devonian, Tafilalt, Morocco.

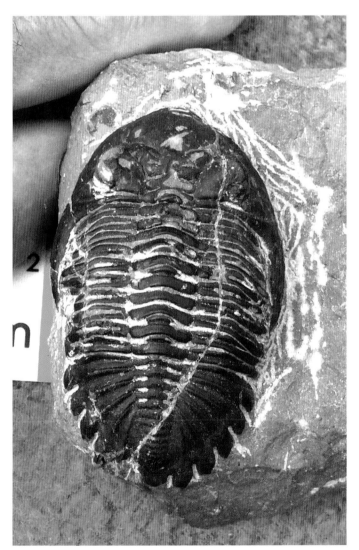

Fig. 06-62. *Metacanthina barrandei*, Oehlert. One of the most common Devonian trilobites from Morocco (and elsewhere), *Metacanthina* is also found to the north in Europe in limestone from the same seaway that deposited the limestone containing this specimen. (Value range F).

Fig. 06-64. *Greenops boothi*. A Middle Devonian trilobite genus similar to *Metacanthina*, but characteristic of the eastern part of North America. Arkona Shale, Middle Devonian, Arkona, Ontario, Canada. (Value range F).

110

Fig. 06-65. *Phacops rana*. A group of a widely collected and distributed trilobite from Northern Ohio. Silica Shale, Sylvania, Ohio. (Value range F, single complete specimen).

Fig. 06-66. *Palmichnium* sp. A distinctive Devonian trace fossil, possibly made by small soft bodied arthropods (trilobitomorphs). Lower Devonian, southwestern New York State,

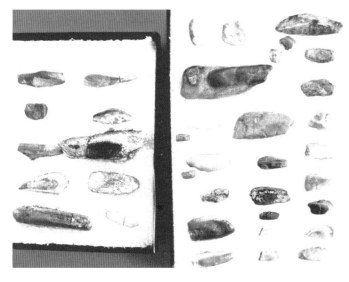

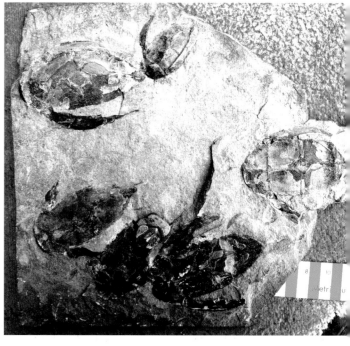

Fig. 06-68. *Bothriolepis canadensis*. A group of primitive placoderms of a type known as an antiarch. Remains of these bony armor fish are found in Late Devonian sediments deposited in fresh and brackish water. Such sediments usually filled rift zones which were connected with the opening (and closing) of oceans that predated the Atlantic. Red sandstone, Escuminac Bay, eastern Quebec, Canada.

Fig. 06-67. *Ptyctodus calceolus*, Newberry and Worthen. These are the crushing teeth of an otherwise unknown placoderm. Placoderms were bony armor fish with jaws. Such teeth can be locally abundant in late Devonian sandstone and shale beds surrounding the Ozark Uplift. Jefferson County, Mo. (Value range H, single tooth).

Fig. 06-69. *Bothriolepis* sp. Fragments of bony armor of an antiarch Placoderm. These are some of the most common Late Devonian bony armor fish. The Catskill Delta deposits, in which it is preserved, are freshwater deposits formed from a series of rivers draining into a rift zone formed during the Devonian Period. Duncannon Member, Catskill Formation, Red Hill, Clinton County, Pennsylvania.

Fig. 06-72. *Dipteris valenciennesi*, Segwick and Murchison. A small lungfish from the Devonian Old Red Sandstone of Scotland. The Devonian is sometimes referred to as the "age of fishes," as it was in this period that many classes of fish first appeared in the rock record, such as lungfish and lobe-finned fish. The Old red sandstone is a series of sandstone and shale beds of fresh water origin which occurs extensively in Scotland and has yielded an extensive fish fauna. It is similar to the non-marine Devonian strata of the Canadian Maritimes and also to beds such as that of the Catskill Delta further south (see previous caption). Devonian land plants are associated also with such deposits which were sediments washed into a Devonian rift zone. Black shale beds in the Old Red Sandstone, Achanarras, Caithenes, Scotland. (Value range D).

Fig. 06-70. Close-up of a partial head shield shown in the previous image.

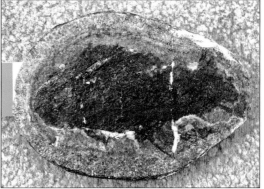

Fig. 06-73. *Cheiraranthus latus*. These early fish were recognized as distinctive fossils early in the history of geology. This specimen is in a concretion. Old Red Sandstone. Gamrie, Scotland. (Value range E).

Fig. 06-71. *Bothriolepis canadensis*. This primitive placoderm (antiarch) is from a locality on the Gaspe Peninsula that has produced a large number of fine Devonian fish. The locality is now a fossil preserve. Escuminac Formation, Miguashua, Quebec. Canada. (Value range D).

Fig. 06-74. *Osteolepis macrolllepidotus*, Agassiz. An example of a small lobe-finned fish. The lobe-fin fish or Crossopterygians was one of a number of fish classes that first appeared in the Devonian. Crossopterygians are considered to be the group which gave rise to amphibians. Sandwick fish bed, Orkney's (Island), Scotland.

Fig. 06-76. *Scaumenacia*. Escuminac Formation, Upper Devonian, Escuminac Bay, eastern Quebec. (Value range B).

Fig. 06-77. Selection of books by Hugh Miller. Upper right, picture of Hugh Miller. Section of Old Red Sandstone on fold out at bottom left, Old Red Sandstone frontispiece at right. Hugh Miller started out as a Scottish mason in the early 19th century where his encounter with fossils in doing masonry work led eventually to his writing a series of books on popular geology and paleontology at the time when strata and associated fossils were first being unraveled. Miller's books sold well in both the U.S. and in Europe. He supported the new field of geology and went against biblical literalism, which was upset by the discoveries of geology. His best known work was *The Old Red Sandstone*. (Value range, for books E).

Fig. 06-75. *Scaumenacia* sp. A nice dipnoan or lungfish from Upper Devonian, fresh water deposits of Escuminac Bay, Illustrations are from a 19th century publication by the British Museum showing similar fossils on display at that institution. Escuminac Formation, Escuminac, eastern Quebec. (Value range B).

Fig. 06-78. *Thursius phdidotus,* Traquair. A crossopterygian or lobe-finned fish. Penryland fish beds, Middle Old Red Sandstone, Thurse, West shore, Caithness, Scotland. (Value range D).

Fig. 06-79. *Eustenopterion* sp. A lobe-finned fish considered as a candidate from which the amphibians evolved. Escuminac Formation, Escuminac Bay, Quebec. (Value range B).

Fig. 06-80. *Eustenopterion*. The characteristically shaped homocercal tail on this specimen is distorted. *Eustenopterion* is an example of a Crossopterygian, believed to be the link between fish and land animals (tetrapods). Escuminac Bay, Quebec. (Value range C).

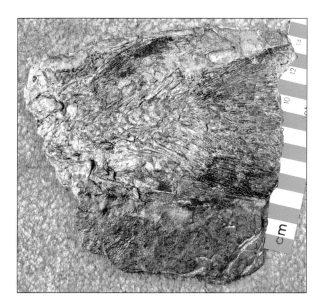

Fig. 06-81. Close up of tail and part of one of the lobe-fins of *Eustenopterion*. Escuminac Formation, Escuminac Bay, Quebec, Canada.

Fig. 06-82. A group of some of the above fish living in fresh or brackish water representative of the Late Devonian environment now "frozen in stone" in the sea cliffs of Escuminac Bay, eastern Quebec. At the bottom right is the placoderm *Bothriolepis*, above it is the lobe-finned fish *Eustenopterion*, and to the left is a group of the lung fish *Scaumrnacia*. Artwork by William E. Brownfield.

Bibliography

Miller, Hugh. *The Old Red Sandstone or New Walks in an Old Field*. Gould and Lincoln, Boston, 1857
_____. *The Testimony of the Rocks or Geology in its Bearings on the Two Theologies, Natural and Revealed*. Gould and Lincoln, Boston, 1857.
Maisey, John G. *Discovering Fossil Fishes*. Westview Press, 1996.

Chapter Seven

The Mississippian Period

The Lower Carboniferous of Europe

The Mississippian or Lower Carboniferous saw the expansion of land vegetation, amphibians and echinoderms.

Seaweed

Fig. 07-001. Fucoid or brown algae(?) "Sea weeds" can leave fossil traces although these are generally vague. This "sea weed" impression is clear and is therefore desirable. It is probably from what are known as the brown algae, one of the common sea weeds living today. St. Louis Formation, St. Louis, Missouri. (Value range F).

Fig. 07-003. *Lepidodendron* sp. Lower Carboniferous (Mississippian) strata of the Canadian Maritimes are almost identical to similar age strata of Europe, particularly in Scotland and Wales. This similarity was first noted by 19th century geologic travelers such as Charles Lyell (1797-1872). In Europe, Mississippian age strata is referred to as Lower Carboniferous, where in much of Europe, it lacks the thick limestone beds like the Burlington and Keokuk Formations of the U.S. midwest or the Red Wall Limestone of the U.S. Southwest. This specimen is from fine grained, "dirty" sandstone which crops out near St. John, New Brunswick, Canada. (Value range F).

Lycopodophyta

These are various types of scale trees or members of the Division Lycopodophyta. They were plants covered by elongated leaves that leave a pattern of distinctive leaf scars.

Fig. 07-004. *Lepidodendron* sp. Lepidodendron stem impressions are often found preserved as compressions in sandstone. This is a stem impression from Upper Mississippian (Chester age) strata near Chester, Illinois. (Value range F).

Fig. 07-002. *Lepidodendron* sp. Scale tree impressions are found in both Mississippian and Pennsylvanian (Carboniferous) strata. This one is particularly clear in its showing the leaf scars that completely covered the trees surface. The so-called "scales" are actually leaf scars produced where the elongate leaves were attached to the plants stem. Sandstone of the Chester Series, Tar Springs Formation, Perry County, Missouri. (Value range F).

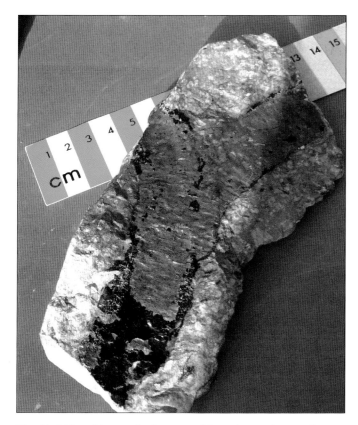

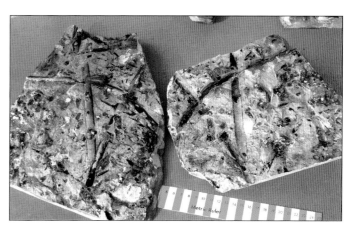

Fig. 07-007. These narrow forms of lycopodophyte are considered to be a different species of *Valdmeyerodendron*. Salem Formation, St. Louis, Missouri (Value range F).

Coral

Corals can be distinctive fossils in rocks formed in a marine environment.

Fig. 07-005. *Valdmeyerodendron* sp. This scale tree impression shows branching stems of this distinctive group of plants belonging to the Lycopodophyta, which today are represented by small, herbaceous plants. This type of "scale tree" is associated with silty limestone beds of the U.S. midwest. It was not as robust as were the other, better known scale trees like *Lepidodendron*. It is named after Valdmeyer, Illinois, in southwestern Illinois. Salem Formation, St. Clair County, Illinois. (Value range F).

Fig. 07-006. *Valdmeyerodendron* sp. This "reedy" scale tree has left a thin film of coal-like material on this part and counterpart. Salem Formation, St. Louis, Mo. (Value range F).

Fig. 07-008. *Chonophyllum* sp. This type of coral is unusual for the late Paleozoic. It is preserved in chert. Sometimes the fossils found in chert are different than those found in associated, similarly aged limestone and can also be much more obvious. Many of the corals found in Mississippian age cherts of the U.S. midwest are closely related to those of the Devonian, but are still different enough to be Mississippian, like this one which is undescribed (that is, it has not been formally recognized and entered into the scientific literature). Burlington Formation, Pike County, Missouri. (Value range F).

Fig. 07-009. *Lithostrotionella castelnaui*. This colonial coral can be locally abundant in the St. Louis Limestone and its correlatives in North America. Here is a specimen placed in a masonry wall. People sometimes think these are petrified wasp nests.

Fig. 07-010. *Lithostrotionella* sp. This is a common coral in Mississippian strata of the U.S. midwest. A somewhat similar coral is found in black limestone of similar age in England and is known under the genus *Lonsdalia*. St. Louis Formation, Middle Mississippian, St. Louis, Mo. (Value range F).

Fig. 07-012. *Syringopora* sp. A slice through a colony of this "pipe organ" coral. *Syringopora* usually occurs in clumps and is a common Mississippian or Lower Carboniferous fossil in the northern hemisphere. Keokuk Formation, Callaway Co., Mo. (Value range G, for polished slab).

Fig. 07-013. *Syringopora* sp. This widespread Mississippian coral can form reef-like clumps like this specimen weathered out from hard limestone in an arctic climate. *Syringopora* is also known as the "pipe organ" coral as its corallites can resemble the vertical pipes of a pipe organ in side view. Middle Mississippian Limestone, Brooks Range, Alaska. (Value range F).

Fig. 07-011. *Lithostrotionella* sp. A single silicified (quartz replaced) specimen of this common "honeycomb" coral. St. Louis Limestone, St. Louis, Mo. (Value range F).

Fig. 07-014. *Amplexus* sp. This solitary coral is a distinctive Mississippian fossil that can resemble a Calamite, which is of course, a fossil plant. This group of specimens came from Mississippian age chert of the Missouri Ozarks, however *Amplexus* is found world wide in similarly aged rocks. Keokuk Formation, Callaway Co., Missouri. (Value range F, single specimen).

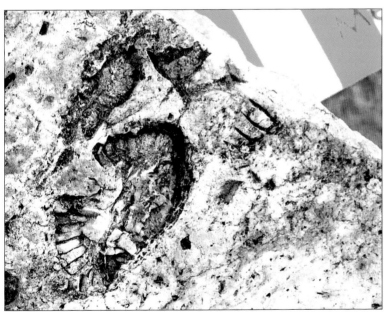

Fig. 07-015. *Amplexus* sp. A cross-section of this typical Mississippian fossil coral, occurring now in the Arctic. When it was living, it was associated with shallow, tropical seas. Much of the Paleozoic strata and fossils of Alaska represent autochronous terrain, that is underlying rock strata of major parts of the state are made up of strata formed somewhere else and transported thousands of miles by sea floor spreading. Mississippian strata formed elsewhere was then incorporated into the crust, in this case northern Alaska. This form of *Amplexus* occurs in chert which resembles that of the same age in other parts of the world. Similar Mississippian strata occurs in the Urals of Russia and this is likely the original source of this fossil. Brooks Range, northern Alaska. (Value range G).

Fig. 07-016. *Hadrophyllum* sp. These silicified "button corals" weather out from Mississippian age limestones exposed in the western Ozarks. Pierson Formation, Fristoe, Missouri. (Value range G for four).

Fig. 07-017. *Neozaphrentis* sp. Silicified specimens of these corals can weather from Mississippian limestones in quantity. They often are picked out of and associated with red clay, which is a weathering product of the limestone. These corals have been informally referred to as "pecker corals," where they can be biostratigraphic markers in the western Ozarks of Missouri, Arkansas, and Oklahoma. Pierson Formation, Fristoe, Missouri. (Value range G).

Conularia sp.

Fig. 07-018. *Conularia missourensis*. Conularids are problematic fossils. Their "shell" or exoskeleton originally was composed of a polysaccharide known as chiton, which is the same material which makes up an insect exoskeleton and is chemically similar to cellulose. Few invertebrate animals other than insects have their hard parts composed of chiton and the quadrilateral symmetry of the fossil is odd. Conularids are sometimes found clustered together and apparently lived upright on the sea floor. They have variously been considered as mollusks, as cnidarians, and even as representatives of an extinct animal phylum. Salem Formation, Godfrey, Illinois. (Value range F).

Fig. 07-019. *Conularia missourensis*. Another group of these peculiar, problematic fossils.

Bryozoans

Bryozoans are varied and interesting marine fossils
in Mississippian rocks.

Fig. 07-020. *Fenestrella* sp. This
shows all of the "cells" or zooecia
characteristic of the bryozoan animal.
The individual bryozoan animal was
quite small and bryozoan fossils like
this represent a colony of individuals.
Bryozoans are not related to corals
(which they resemble) but are related
to brachiopods (lophophorates).
The name fenestrella refers to the
openings or "cells" occupied by the
animal as "little windows" (Latin-
fenestrae=windows) which the
bryozoan zooecia resemble.

Fig. 07-021. This bryozoan colony is shaped
like a lyre with the bryozoan animals zooecia
forming the "strings" of the lyre. Such zooecia
resemble Spanish lace. Chester Series, Upper
Mississippian. (Value range G).

Fig. 07-022. *Evactinopora* sp. This
genus of radial symmetrical bryozoans
is found in Mississippian strata of the
U.S. midwest. Bryozoans of the late
Paleozoic produced an axis (possibly in
conjunction with some other life form
such as an algae) consisting of interesting
and pleasing fundamental geometric
shapes. Burlington Formation,
Hannibal, Missouri (Value range F).

Fig. 07-023. *Evactinopora sexradiata.* Five or six branches
characterize this small species of the bryozoan genus
Evactinopora. Fern Glen Formation, Lower Mississippian,
Jefferson Co., Mo. (Value range F).

Fig. 07-024. *Archimedes wortheni*, Hall. The corkscrew-shaped, central
axis of this distinctive bryozoan is a relatively common fossil in Middle
Mississippian Limestones of the U.S. midwest. It has been reported
elsewhere, but is rare in areas other than the United States. The
fenestrellate (Spanish lace) zooecia radiated out from its spiral axis
which some paleontologists believe was produced by a red algae and
Archimedes is an example of fossil symbiosis between an algae and a
bryozoan. Warsaw Formation, Lincoln Co., Mo. (Value range F).

Fig. 07-025. *Archimedes wortheni,* Hall. A large specimen of this distinctive bryozoan. *Archimedes* may be an example of biological symbiosis in its being a cooperative arrangement between a bryozoan colony and a calcareous red algae to which the central axis of the fossil has some structural resemblance. Warsaw Formation, Middle Mississippian (Value range E).

Fig. 07-026. *Archimedes communicus.* A species of *Archimedes* characteristic of the Upper Mississippian (Chesterian strata) of the midwest and southern U.S.

Brachiopods

Brachiopods are characteristic Mississippian fossils.

Fig. 07-027. *Spirifera grimesi,* Hall. These large brachiopods are fairly common in Mississippian limestone of the U.S. midwest. Brachiopod fossils in Paleozoic rocks can be abundant and exist with a great amount of species diversity. They are certainly one of the most common and diverse Paleozoic fossils, contrasting with today where they are minor faunal elements in modern oceans. Only a few examples of this large group are shown here; what is presented is very minimal for this large phylum, however more technical works such as *Index Fossils of North America* or the *Treatise on Invertebrate* paleontology (see index) can enable one to more thoroughly delve into them

Fig. 07-028. *Orthotetes keokuki.* These flat and thin brachiopods can be locally common in the U.S. midwest where they often grouped together on the shallow sea floor. Keokuk Formation, Middle Mississippian. (Value range F, single specimen).

Blastoids

Blastoids are an extinct class of echinoderms that are abundant in Mississippian strata of North America.

Fig. 07-030. *Globoblastus* sp. These silicified external and internal steinkerns of these globular blastoids are preserved in chert, a common fossil preserving medium in Mississippian age rocks in many parts of the world, but particularly in the U.S. midwest. Burlington Formation, Middle Mississippian. (Value range F).

Fig. 07-031. *Orbitremites* sp. These are one of the abundant blastoids found in thick, Middle Mississippian cherty limestones of the U.S. midwest. (Value range G, single specimen).

Fig. 07-032. *Pentremites* sp. A blastoid with both stem and pinnules preserved. Blastoids are usually found as specimens lacking both stem and pinnules. Blastoids with the stem are rare. The pinnules, which were involved in filter feeding, are also rarely preserved. Glen Dean Formation, Sulphur, Indiana. (Value range E).

Fig. 07-033. *Pentremites* sp. Another blastoid with the stem preserved. Blastoid stems are narrower, smaller, and more delicate than were the stems of contemporary crinoids. Glen Dean Formation, Chester Series, Anna, Illinois. (Value range F).

Fig. 07-034. *Pentremites globosus.* A cluster of blastoids from Chesterian age strata (Upper Mississippian) from southeast Missouri. (Value range E).

Opposite
Fig. 07-029. Left: *Spirifer;* right: *Schizophoria* sp. Examples of these two common Mississippian brachiopods, which are preserved as internal chert steinkerns (molds). Brachiopods are the largest or nearly the largest group of Paleozoic megafossils. Brachiopod species diversity reached its zenith during the late Paleozoic. Burlington Formation, Callaway Co., Mo. (Value range G for both).

122

Fig. 07-035. *Pentremites* sp. These blastoids come from a locality in southwestern Illinois, which has produced thousands of specimens where they occur in a blueish-grey shale bed. Paint Creek Formation, Chester Series. Prairie du Long Creek, St. Clair Co., Illinois. (Value range H, single specimen).

Fig. 07-036. *Pentremites globosus.* These large blastoids occur in southern Illinois as well as in Kentucky and Tennessee. They are some of the largest blastoids found anywhere. They are very nice! Golconda Formation Chester Series, southern Illinois. (Value range F, single specimen).

Fig. 07-037. Another group of large Upper Mississippian blastoids with a plate from a 19th century work describing them.

Edrioasteroids

Edrioasteroids are another extinct class of Paleozoic echinoderms.

Fig. 07-038. *Discocystis kaskaskiensis*, Hall. Edrioasteroids are relatively rare (and therefore desirable) fossils. This relatively large genus, comes from Upper Mississippian age strata (Chesterian) of the U.S. southeast. Paint Creek Formation, Chester Group, Prairie du Long Creek, southwest Illinois.

Echinoids

Echinoids or sea urchins were successful echinoderms during the Mississippian Period.

Fig. 07-040. *Echinocrinus (Archeocidarus)* sp. Scattered plates and spines of this echinoid can be relatively common fossils in late Paleozoic rock strata but complete specimens are quite rare. This specimen, a reconstruction of two Archeocidarid specimens gathered from a concentration of plates and spines. St. Louis Limestone, Arnold, Missouri. (Value range F).

Fig. 07-041. *Echinocrinus (Archeocidaris) shumardiana,* Hall. These are impressions in chert of an archeocedarid sea urchin. Such fragmental material is what is usually found with these early and delicate sea urchins. Keokuk Formation, Paris, Missouri. (Value range F).

Fig. 07-042. *Echinocrinus (Archeocidaris)* sp., with a plate from an early 20th century monograph illustrating similar specimens. (Value range G).

Fig. 07-043. *Echinocrinus (archeocidaris)* sp. A slab of somewhat disarticulated specimens with the spines being predominant, a common occurrence with this early urchin.St. Louis Limestone, St, Louis, Missouri. (Value range G).

Fig. 07-044. *Oligoporus missourensis,* Meek and Worthin. A small, chert steinkern of one of the most desirable Paleozoic sea urchins (echinoids). These are the oldest fossil echinoids that can normally be found complete. They had a test made up of robust, interlocking plates. Specimens are found as internal molds in chert as is the cast here. Burlington Formation, Middle Mississippian, Eureka, Missouri. (Value range E).

Opposite
Fig. 07-039. Echinocrinus (Archeocidarus) agassizi. Cidarids are sea urchins that became well established in the Mesozoic Erawhich and live today. Mesozoic specimens of the genus *Cidarus* are to the right. *Archeocidarus* is a Paleozoic ancestor to *Cidarus.* The plates of *Archeocidarus* were held loosely together so that complete specimens are almost never found. They occur either as scattered plates and spines or as flattened tests with the spines scattered. St. Louis Limestone, St. Louis, Missouri. (Value range F).

Fig. 07-045. *Oligoporus* sp. This is an adult, normal size *Oligoporus* specimen preserved in chert as a chert steinkern. Specimens of this sea urchin are found in the U.S. midwest, but their uniqueness and rarity makes them desirable fossils. They are rather distinctive looking and sometimes are found in unlikely places such as garage sales and jumble shops. Persons who are "rock conscious" might pick up one of these as they really look different, and they may also occur as an adjunct to an artifact collection. This specimen came from a farmer in Arkansas who had placed it, along with other rocks of interest, around a tree in the front yard of his house. Boone Formation, Jasper County, Arkansas. (Value range D or E).

Fig. 07-04. *Lovenechinus missourensis*, Jackson. A small specimen of this desirable early urchin preserved as a chert steinkern. Illustrated next to the fossil is part of a plate of illustrations from an early 20th century monographic work on early urchins. The chert specimens illustrated in this work came from a lead-zinc mine near Webb City in southwestern, Missouri, where a cluster of large specimens was found by miners around 1899. The specimens from this mine are beautifully covered with a fine druse of sphalerite crystals making them very striking as well as beautiful mineral specimens. Keokuk Formation, with illustration of some of the Webb City specimens.

Fig. 07-046. *Oligoporus* sp. A chert steinkern (internal mold) of this large sea urchin from a zinc mine near Webb City, Missouri. A pocket of these was found in a mine in southwest Missouri in 1899. (Value range D). Courtesy of Washington University, Dept. of Earth and Planetary Sciences collection.

Fig. 07-048. *Melonechinus (Melonites) multiporus*, Norwood and Owen. This is the St. Louis, Missouri fossil urchin. This desirable sea urchin is found in the St. Louis region as well as possibly elsewhere where fossiliferous beds of the St. Louis Limestone outcrop. Specimens of this sea urchin have been widely distributed, particularly toward the end of the 19th century to museums and institutional fossil collections. Individual plates of this urchin are rather common in the upper part of the St. Louis Limestone, but complete specimens are rare. (Value range D).

Fig. 07-049. *Melonechinus* sp. A large, flattened specimen of the primitive(?) sea urchin *Melonechinus*, which shows the small spines that covered the globular-shaped urchin. The urchin was the size and shape of a grapefruit. (Value range D).

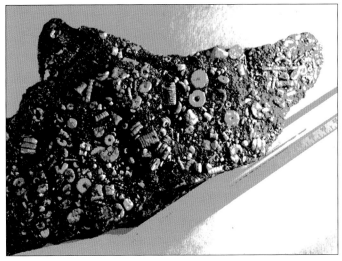

Fig. 07-051. Crinoidal limestone. The Burlington and Keokuk formations of Missouri, Illinois, and Iowa can have beds composed primarily of crinoid fragments like these. Native Americans chose to gather the circular stem fragments and string them as beads.

Crinoids

Crinoids were very dominant and successful echinoderms during the Mississippian. They are also desirable fossils when complete.

Fig. 07-050. Part of a crinoid holdfast, the structure that anchors a crinoid to the sea floor. Such pieces of crinoid stems and holdfasts can be common fossils in many parts of the world. This specimen is of historic interest in that it came from the collection of Dr. J. P. Yandell of Louisville, Kentucky. Dr. Yandell collected fossils and became interested in geology in the 1830s, when modern geology, with its emphasis on megatime, came on the scene. He collected fossils in the Louisville area from the 1830s to the Civil War when the geology of the U.S. midwest was first being unraveled. Dr. Yandell was also a close friend of David D. Owen of New Harmony, Indiana, as well as of Charles Lyell of London, who visited with him in the spring of 1848 on Lyell's second North American geology trip. One of the places Dr. Yandell took Lyell to collect fossils was Button Mold Knob in Louisville, where Dr. Yandell collected this crinoid holdfast.

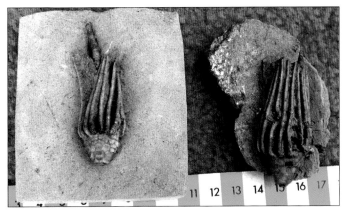

Fig. 07-052. *Macrocrinus mundulus*. A phenomenal number of complete, high quality crinoids have been collected from Mississippian age calcareous siltstones which crop out in the vicinity of Crawfordsville, Indiana. First discovered and worked in the mid-19th century, Crawfordsville crinoids are some of the most complete and beautiful in the world. This is one of the more commonly found ones in the rather large fauna. Ramp Creek Formation, Borden Group, Osagian Series, Indian Creek locality. (Value range F, single specimen).

126

Fig. 07-053. Left: *Scytalocrinus robustus;* right: *Onychocrinus ulrichi.* This group of crinoids is representative of the richness of the Crawfordsville locality. Numerous slabs of these crinoids were obtained in the 1980s and 1990s for the fossil market. These came from both the Indian Creek locality, south of Crawfordsville, and from the original Crawfordsville locality along Sugar Creek. Ramp Creek Formation, Borden Group, Crawfordsville, Indiana.

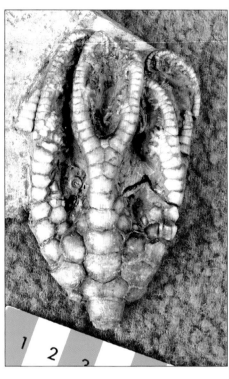

Fig.07-055. *Onychocrinus exculptus.* A distinctive crinoid recently collected (1985) from the Crawfordsville, Sugar Creek locality. Edwardsville Formation, Borden Group, Crawfordsville, Indiana. (Value range F).

Fig. 07-054. *Taxocrinus colletti,* White. Two specimens of this distinctive crinoid of the Taxocrinoidea, a Paleozoic crinoid order well represented by the Crawfordsville fossils. Ramp Creek Formation, Borden Group 1, Indian Creek, Indiana.

Fig. 07-056. *Abrotocrinus* sp. Two specimens from the Crawfordsville Sugar Creek locality collected near the end of the 19th century. Such specimens from old collections are much darker in color than are those collected more recently; the dark ones are genuine "antique" fossils and do have an old look about them. Other "antique fossils," such as the British ammonites collected in the early 19th century and found in old collections, also have this antique look about them. Edwardsville Formation, Borden Group, Sugar Creek near Crawfordsville, Indiana. (Value range F, single specimen).

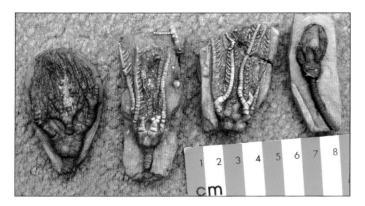

Fig. 07-057. *Barycrinus* sp. (left) A small specimen plus other crinoids (*Abrotocrinus*) collected in the late 19ᵗʰ century from the Sugar Creek, Crawfordsville locality. (Value range F).

Fig. 07-058. *Platycrinites* sp. A relative large crinoid with large basal plates on its calyx. Ramp Creek Formation, Borden Group, Indian Creek, Indiana. (Value range E).

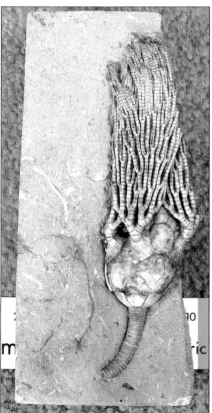

Fig. 07-059. *Platycrinites* sp. A spectacular specimen of this attractive crinoid with its large basal plates. Indian Creek Locality, Borden Group. (Value range E).

Fig. 07-060. *Halysiocrinus nodosus*, Hall. One of a number of crinoids that resemble a 1940s vacuum cleaner. Ramp Creek Formation, Borden Group, Waveland, Indiana. (Value range E).

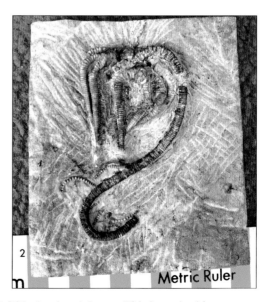

Fig. 07-061. *Agaricocrinites* sp. This is a crinoid genus characteristic of the Late Osagian crinoid "gardens." The Osagian (Lower Middle Mississippian) was the peak of crinoid diversity in North America, where some of the most extensive crinoid faunas occur. Ramp Creek Formation, Indian Creek, Waveland, Indiana.

Fig. 07-062. *Taxocrinus colletti*, White. A typical quality specimen of this robust crinoid from the Crawfordsville, Sugar Creek locality. (Value range E).

Fig. 07-065. *Culmicrinus missourensis. Culmicrinus* was a robust and relatively thick-stemmed crinoid of the Meramecian (Late Middle Mississippian). St. Louis Limestone, St. Louis, Missouri.

Fig. 07-063. *Cactocrinus imperator.* An attractive crinoid with a fine set of pinnules from Mississippian limestones of northwestern Iowa. The age of this limestone and its associated fossils is debated. Its crinoid fauna is totally different from that of Mississippian age limestones of the southeastern part of the state. Gilmore City Formation, Gilmore City, Iowa (Value range E).

Fig. 07-064. *Culmicrinus missourensis*. A distinctive, but rather plain looking crinoid from the Middle Mississippian, St. Louis Limestone of St. Louis, Missouri. (Value range E).

Fig. 07-066. *Platycrinites* sp. A crinoid calyx preserved as an internal mold or chert steinkern from Mississippian age outliers of the Missouri Ozarks. Localized occurrences of mid- and late Paleozoic rocks occur over the Ozarks where they often cap the tops of knobs (small mountains) in the Ozark Uplift of Missouri and Arkansas. The fossils associated with such knobs and outliers (which are remnants of a once more continuous existence of younger strata) are sometimes different from the those found in similar age strata away from the Ozarks. There is also a tendency for these Ozark outlier fossil faunas to be rich in mollusks, as the waters of these shallow seas were of greater salinity than were areas distance from the uplift, and mollusks can tolerate higher salinity better than many other animal phyla. Osagian Outlier (Lower Middle Mississippian), Rolla, Missouri. (Value range F).

Fig. 07-067. A delicate crinoid calyx preserved as an external and internal mold (steinkern) in yellow, iron stained chert. Mississippian age chert in parts of midwest North America can yield such fossils, when the calcite plates of the crinoid have been dissolved away leaving an impression of the crinoid head or calyx in the chert. Such chert internal and external molds can be quite delicate and detailed. Burlington Formation, Fulton, Missouri. (Value range E).

Fig. 07-069. *Batocrinus* sp. Two chert steinkerns of this relative common crinoid from the Burlington Formation.

Fig. 07-068. Close-up view of the same specimen as in the previous image, showing the internal mold of the crinoids calyx. Such chert steinkerns are quite attractive but they are difficult to identify as to genus and species. (Value range E).

130

Fig. 07-070. Group of Burlington crinoids. The Burlington Limestone of the U.S. midwest represents the deposits of a shallow, epicontinental sea. There is a considerable diversity in crinoid genera and species which lived in these "crinoid gardens," making up what is one of the largest crinoid faunas (fossil or living) in the world.

Fig. 07-071. Another group of crinoid calyces illustrative of the richness of Burlington crinoids found in the U.S. midwest. The Burlington Formation covers parts of Iowa, Illinois, Missouri, Arkansas (Boone Formation), and Oklahoma. For the crinoids to show up, the rock has to be deeply weathered. Usually under soil, fresh limestone shows nothing. These specimens are all from the Springfield, Missouri, area where deeply weathered Burlington Limestone makes up parts of the Western Ozarks. (Value range, individual specimen G).

Fig. 07-073. *Platycrinites* sp. This is a very elongate form of *Platycrinites*. Burlington Limestone, Springfield, Missouri. (Value range E).

Fig. 07-072. Burlington or Osagian "crinoid garden" seascape. Clear, warm waters of the shallow, low latitude seaways that covered parts of southern North America teemed with life, especially crinoids. Oil painting by Virginia M. Stinchcomb.

Fig. 07-075. Group of Keokuk crinoids. A group of crinoids from the Upper Osagian Keokuk Limestone or Formation, part of a thick sequence of limestone with the Burlington Limestone below it. Both are rich in crinoids. (Value range F).

Fig. 07-074. *Uperocrinus pyriformis*, Shumard. The arms and a part of the tegmen (at upper left) are features seen in this specimen that usually are not preserved on Burlington crinoids. The author of the species, Benjamin Shumard, was a pioneer American geologist-paleontologist during the mid-19th century, when the rich fossil resources of the U.S. midwest were initially being documented. Burlington Limestone, Lincoln County, Missouri. (Value range E).

Fig. 07-077. *Doryocrinus mississippensis*, Roemer. Specimens of this crinoid genus have five large spines which project from the calyx. Three spines are on the middle specimen in the picture, the other two are lacking the spines. Keokuk Formation, Jerseyville, Illinois. (Value range E).

Fig. 07-076. Group of Keokuk crinoids. Another group of different crinoids from the Late Osagian Keokuk Formation of the U.S. midwest.

132

Mollusks

Mollusks, particularly gastropods and cephalopods, were successful organisms in Mississippian seas.

Fig. 07-078. *Bellerophont scissile.* A large, bellerophont gastropod preserved in red jasper. Many Mississippian fossils, particularly those of the western U.S., are replaced with brick-red jasper. The bellerophontids represent a family of planispiral gastropods which are restricted (except for a few in the Triassic) to the Paleozoic Era. Ste. Genevieve Formation, Ste. Genevieve, Missouri. (Value range E).

Fig. 07-079. *Platyceras* sp. These are fairly common gastropods in Mississippian rocks, but their diversity in shape can be frustrating and overwhelming, particularly with some occurrences. *Platyceras* is unquestioned as to its gastropod affinity, however the novice can mistake them for monoplacophorans, which some forms of *Platyceras* resemble. *Platyceras* is a peculiar gastropod that can be almost bilaterally symmetrical on many (but not all) specimens. This seems to be particularly true with those found in groups in chert concretions and with some specimens that were attached to crinoids. Specimens at the top from shale beds of the Warsaw Formation; those at the bottom from cherts of the Keokuk Formation.

Fig. 07-080. *Platyceras* sp. These chert steinkerns from the Keokuk Formation represent a type of *Platyceras* associated with the clear seas of the "platform" limestones of the U.S. midwest. Specimens of *Platyceras* associated with muddy sediments, like shale and siltstone, are often larger and may be associated with the tegmen of crinoids (corprogenic feeding). Many (most) of the late Paleozoic fossils of North America have been well studied and documented, however there is still a plethora of different forms of *Platyceras*, particularly in Mississippian chert whose specimens have not been "sorted out" and documented in the scientific literature.

Fig. 07-081. *Platyceras* sp. These are broad forms of this peculiar gastropod that gastropod worker Ellis Yochelson considered to be "frustratingly diverse and difficult in their morphologic diversity." The different shell shapes may all be one species, but again maybe they're not! *Platyceras* seems to occur more commonly as chert steinkerns than in associated limestone. Sometimes they are clustered together in a chert chunk suggesting that they were gregarious. *Platyceras* that were associated with muddy environments are often associated with crinoid tegmens to which they were apparently attached.

Fig. 07-082. *Euomphalus (Straparolus) latus*. These relatively large gastropods come from isolated remnants of once more widespread Mississippian strata in parts of the Ozarks known as outliers. Such outliers have a somewhat different fossil fauna than that of similar age rocks formed off of the Ozark Dome. In the case of late Paleozoic outliers, seas which deposited the strata of the outliers were more saline than normal, a condition which could be better tolerated by mollusks than by other marine invertebrates. Osagian outliers, Rolla, Missouri (Value range F, single specimen).

Fig. 07-083. *Rayonnoceras solidiforme*, Croneis, 1926. A relatively large, straight cephalopod that lived on a murky, organic rich sea bottom. This organic rich sediment became the Fayetteville Shale which yields *Rayonnoceras* when these black, petroleum-rich shales weather or are exposed in streams. The massive sections of the cephalopod which are found are usually the highly mineralized part of very large shells, some of which were seven to ten feet in length. Fayetteville Shale, Upper Mississippian, northwestern Arkansas. (Value range F, single specimen).

Fig. 07-084. Seascape of organic rich, and relatively deep seas with the large cephalopod *Rayonnoceras*. Artwork by Virginia M. Stinchcomb.

Trilobites

Trilobites were on the decline in the Mississippian. They are found less frequently in Mississippian marine strata than in that of the Devonian.

Fig. 07-085. *Phillipsia* sp. A complete, enrolled trilobite preserved as a chert steinkern from Osagian age outliers near Rolla, Missouri. Such complete trilobites, from unusual geologic environments such as this, are rare and therefore desirable. Osagian outliers, Rolla, Missouri, (Value range E).

Fig. 07-086. *Phillipsia*. A group of pygidia from Osagian outliers, Rolla, Missouri. The pygidia is the most conspicuous part of late Paleozoic trilobites and is the part most commonly collected. (Value range G, single specimen).

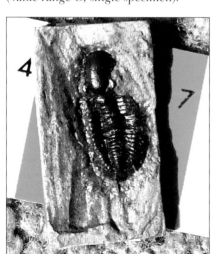

Fig. 07-087. *Kaskia chesterensis*. Most late Paleozoic trilobites are small and relatively rare, although this specimen is missing its fixigenia (free cheeks). Chester Series, Alabama. (Value range F).

Fig. 07-088. *Breviphillipsia sampsoni*. A number of these small, but complete Lower Mississippian trilobites have come onto the fossil market from a locality near Glasgow, Missouri. They are some of the more readily available late Paleozoic trilobites. Compton Formation, Chouteau Group, Kinderhookian (Lower Mississippian) Series. Glasgow, Missouri. (Value range F).

Shrimp

Shrimp are arthropods of the class Crustacea.

Shark-like Fish

These are the teeth and spines of shark-like fishes.

Fig. 07-090. *Sphenacanthus (Ctenacanthus)* sp. Sharks or shark-like fishes lived in clear, shallow Mississippian seas. They are primarily known from their teeth and their dorsal spines such as this. These spines are referred to as Ctenacanthid spines after *Ctenacanthus*, a genus established in the 19th century and, as is the case with most early established genera, later split by paleontologists into a number of "new" genera. The bearer of Ctenacanthid spines did not seem to be directly related to modern sharks, which appeared in abundance in the late Mesozoic (Cretaceous Period). Burlington Limestone, Hannibal, Missouri. (Value range F).

Fig. 07-091. *Ctenacanthus* occurs sporadically in Mississippian age limestones of the U.S. midwest and can occur there with a considerable amount of size and morphologic diversity. *Ctenacanthus* spines can be large enough to attract the notice of "marginally interested" fossil collectors. The author has seen fragments of specimens that must have been over a meter in length if they had been complete. Such large specimens are desirable both scientifically and monetarily. Specimen at left is in chert, an uncommon occurrence for a vertebrate. St. Louis Limestone, St. Louis, Missouri. (Value range highly variable with large, well preserved specimens being in the B-C range, but such monetary consideration should be superseded by scientific considerations).

Fig. 07-092. *Chomatodus* sp. A number of genera and species have been established to accommodate the teeth of Paleozoic shark-like fishes that are also known by the general term of bradyodonts. *Chomatodus* is one of the more peculiar forms. Tooth at upper right is not *Chomatodus* sp., but *Deltodus* sp. St. Louis Limestone, St. Louis, Mo. (Value range G, single specimen).

Opposite
Fig. 07-089. Compressions of these relatively small fossil shrimp cover bedding planes in a horizon of brackish water shale near Edinburgh, Scotland. The shale beds in which they occur are similar to Mississippian oil shale which occurs in New Brunswick, Canada, and may be part of the same lake deposit. Associated with these fossil shrimp have been found compressions of the conodont animal. Conodonts are small, microscopic tooth-like fossils widely distributed in Paleozoic rocks and quite useful in biostratigraphy. They are now believed to be primitive vertebrates (see Gould, 1985). Granton shrimp beds, Edinburgh Scotland. (Value range G).

136

Fig. 07-093. *Deltodus* sp. *Deltodus* is a flat, deltoid-shaped (shaped like a delta) tooth which is one of the more commonly found Mississippian bradyodont tooth type. Some Deltodus teeth can get fairly large. Keokuk Formation, Lincoln County, Mo. (Value range E).

Fig. 07-094. *Cladotus* sp. Teeth of the genus *Cladotus* look like sharks teeth more than do the teeth of most other bradyodonts. The various tooth types formed the animals "tooth bank" or "tooth pavement" in their mouth which enabled their owners to engage in highly predatory behavior. Presumably the different tooth types represent different forms of shark-like fish, however this is only an inference since few complete or even partially complete assemblages of these teeth have been found. It is hypothesized that these "pavement tooth" sharks may have fed upon crinoids whose calyxs (heads) they crushed and ate. This also helps explain an almost total absence of crinoid heads in some crinoidal limestone. The crinoid calyx shown here (bottom) has a puncture in it which matches the size and shape of a cladodont tooth. All from Keokuk Formation. (Value range F, single specimen).

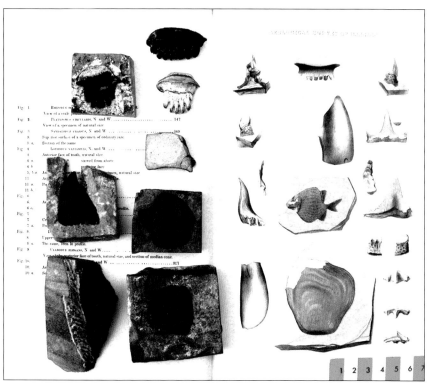

Fig. 07-095. A selection of bradyodont teeth and pictures of such in a late 19th century work on them. Not only are the fossil teeth collectable, but original literature on them, such as this, is desirable. The hand working of limestone beds in the 19th and early 20th century for masonry stone resulted in the recovery of many teeth and spines of shark-like fishes, which are described in these reports by St. John and Worthin, 1875. (19th century works on shark-like fishes, value range E).

Fig. 07-096. *Bythiacanthus sp.* This short, stubby fish spine was originally described in late 19th century works on U.S. midwest fossil fishes.

Fish

Fossil fish such as these are found in ancient lake deposits.

Fig. 07-097. *Rhadinichthys* sp. This is an example of a Paleoniscoid fish, a group of Paleozoic fish also known as ray-finned fish, that predate the more modern bony fish of today. It comes from an outcrop of Mississippian shale which is petroleum rich enough to have been mined in the past. Albert Mines in the 19th century was a source of Albertite, (an asphalt-like product and Ozocerite, a hard, natural mineral wax). The shale was also processed for oil before drilling technology made modern petroleum production feasible, The oil shales of Albert Mines were also a source of many fine fossil fishes like this one. (Value range E).

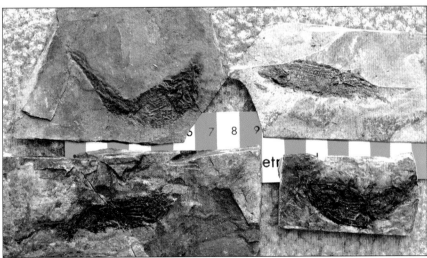

Fig. 07-098. A group of typical paleoniscoid fossil fishes that can be locally found in the Mississippian oil shales. The fish are more primitive and quite different from the fossil fish commonly seen from oil shale deposits in Wyoming, which are much younger (Early Cenozoic). Oil shales, with their fossil fish were generally deposited in lakes formed during periods of tectonic activity, in this case tectonic activity which accompanied the initial formation of the Appalachian Mountains. (Value range F, typical specimen).

Fig. 07-099. A paleoniscoid fish in finely laminated shale, deposited in brackish water from the Ural Mountains and similar to the shales and fish of the Albert Mines in New Brunswick, Canada.

Dipnoans

These are fossil Dipnoans or lung fish.

Fig. 07-100. *Caridosuctor populosom*, a coelacanth. Limey shale deposited in brackish water in what is now central Montana has yielded a number of complete fossil fish, some quite unusual, in addition to yielding some early fossil amphibians. Coelacanths, also known as lobe-fin fish, were known only from fossils until 1938 when a living coelacanth was found in the Indian Ocean. They are believed to represent the link between fish and tetrapods (amphibians) and as such they upset creationists. Coelacanths can live in water respiring by gills, but they can also breath oxygen through a primitive lung. Lung fish, which are often found as fossils with coelacanths, have been candidates for being the transitional animals between fish and amphibians, however coelacanths are the more likely participant in this significant transition. The Bear Gulch shale, is one of the paleontological windows of the late Paleozoic. Bear Gulch Shale, central Montana.

Fig. 07-101. Both dipnoans and coelacanths have been considered as the link between fish and tetrapods (amphibians) and as such they both upset creationists. This coelacanth was initially thought to be a dipnoan (lungfish), as its tail fin has been distorted in such a way that it resembles the heterocercal tail of a lungfish. Bear Gulch Shale, central Montana.

Bibliography

Ausich, William I. "Lower Mississippian Burlington Limestone along the Mississippi River Valley in Iowa, Illinois and Missouri, USA," in *Fossil Crinoids*. Cambridge University Press, Cambridge England, 1999.

_____. "Lower Mississippian Edwardsville Formation at Crawfordsville, Indiana, USA," in *Fossil Crinoids*. Cambridge University Press, Cambridge England, 1999.

Gould, Stephen J. *Reducing Riddles in The Flamingo's Smile, Reflections in Natural History*. W.W. Norton and Company, New York and London, 1985.

Chapter Eight
The Pennsylvanian Period

The Coal Age or Upper Carboniferous

During the Pennsylvanian or Upper Carboniferous, large parts of the earth's land masses were covered with warm, tropical swamps. It was in such swamps that much of the coal that powered the Industrial Revolution in Europe and North America was formed.

Tree Ferns

Tree ferns were the dominant plants in Pennsylvanian coal swamps. Here is some of the tree fern petrified wood and foliage.

Fig. 08-002. *Callipteridium membranaceum*, White. Ferns such as this from Pennsylvanian rocks are often labeled under the genus *Pecopteris* sp. More detailed (or reductionist) work on the plants of the Pennsylvanian coal floras has split many well known genera such as *Pecopteris* into numerous genera and species, Such is the case with this frond from western Missouri. This specimen came from a localized flora of large, complete fronds associated with a thick, localized coal seam mined in the late 19th century. This coal bed and shales overlying it filled a sinkhole formed in the underlying limestone (paleokarst). The exceptional flora associated with this occurrence was described by David White, USGS Monograph 37, 1899. (Value range E)

Fig. 08-001. Tree fern wood. Large petrified logs can be found weathered from Middle Pennsylvanian age rocks in Illinois, Missouri, Tennessee, and Alabama. Slices of these logs, like those here, show no annular rings or other regular structure and are often full of cavities which can be lined with quartz crystals. The logs are generally brown in color and resemble chunks of somewhat rotten driftwood. Such petrified wood is believed to be from large, rambling tree ferns that lived in coal swamps covering the U.S. midwest during the mid-Pennsylvanian. The lack of annular rings indicates a lack of seasonal climate diversity. It was humid year round, a consequence of North America being close to the equator at this time (approx. 300 million years ago). These slices are from the northern Ozarks, Callaway Co., Missouri. (Value range F for polished slice).

Fig. 08-003. *Asterotheca* cf. *arborescens*. A frond of this narrow-leaved pecopterid genus. From roof shales of the Middle Pennsylvanian Crowberg Coal, Henry County, Missouri (Value range E).

Fig. 08-004. Group of "fern" compression fossils from Middle Pennsylvanian strata of Henry County, Missouri. Top left: *Neuropteris* sp.; bottom left: *Pecopteris* sp. Similar groups of ferns have come onto the fossil market through MAPS expo from outcrops near Warrensberg, Missouri, as well as elsewhere. Fern fossils from the eastern half of the U.S. are widespread in their occurrence and as a consequence of this abundance they are common and priced accordingly.

Fig. 08-007. A late Pennsylvanian tree fern genus from siltstone beds. The organic material is gone from this fern impression, the plants presence now marked by yellow iron oxide, a type of preservation commonly found in fossil plants of the Permian Period. A considerable number of these fern fossils in siltstone came from highway construction in the Kansas City, Missouri, area and have been preserved by collectors. Winterset Formation, Bonner Springs Member, Upper Pennsylvanian. Platte County, Mo. (Value range F).

Fig. 08-005. *Pecopteris* sp. Pennsylvanian (Upper Carboniferous) plant floras of Europe are almost identical to those of North America indicating that the coal swamps of North America spread continuously from America to Europe. There was no Atlantic Ocean at that time; it would open and widen in the Mesozoic Era. These pecopterids are from a coal mine in Belgium. (Value range G)

Fig. 08-008. *Alethopteris* sp. Black, slaty shale overlying an anthracite coal bed near St. Clair, Pennsylvania, is packed with these ferns, and large numbers of them have entered the fossil market and the fossil collecting community. They stand out because the fern leaf has been coated with white kaolin which contrasts with the black, slaty shale. St. Clair, eastern Pennsylvania. (Value range G or H depending upon size of slab).

Fig. 08-006. *Alethopteris grandius*. These distinct fern impressions from near Ottawa, Kansas, have had wide distribution. This particular species of *Alethopteris*, with its tough and leathery leaves, is preserved in siltstone, making particularly clear and distinct leaf impressions. Lawrence Shale, Upper Pennsylvanian. Ottawa, Kansas. (Value range F).

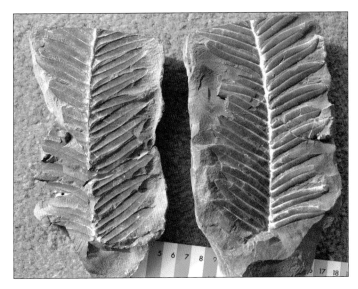

Fig. 08-009. *Alethopteris serli*, Brongniart. A large alethopterid, with its characteristic thick leaves, can form a distinct impression or compression such as found in the larger Braidwood ironstone nodules. The Braidwood or Mazon Creek, Illinois, ironstone nodules along with the marine Essex ironstone nodules, form paleontological "windows" to the Paleozoic Era. These areas have produced tens of thousands of specimens which have had world wide distribution. Francis Creek Shale, Middle Pennsylvanian, Braidwood, Illinois. (Value range F).

Fig. 08-010. A group of typical ferns in ironstone nodules or concretions from coal mine "spoil piles" near Braidwood, Illinois. Upper left: *Neuropteris* sp.; upper middle: *Pecopteris* sp. (part and counterpart); upper right: *Pecopteris* frond; bottom left: *Pecopteris* (part and counterpart); bottom right: *Odonthopteris* sp. (part and counterpart). (Value range F, single specimen, part and counterpart).

Fig. 08-012. *Pecopteris* sp. These specimens of the widespread genus *Pecopteris* are unusual in their being preserved in limestone. Middle Pennsylvanian, central Illinois. (Value range G).

Fig. 08-013. Right: *Neuropteris flexuosa*; left: *Pecopteris* sp. Shale of the Pleasanton Group, Uppermost Middle Pennsylvanian. St. Louis, Co. Missouri, both from localities now covered by urbanization. (Value range G).

Fig. 08-011. Braidwood, Illinois, ferns in ironstone concretions. Top left: *Alethopteris serli*; top right: *Pecopteris* sp., frond, part and counterpart; bottom left: *Alethopteris serli* part and counterpart; right: *Odonthopteris* sp.

Fig. 08-014. *Mixoneura* sp. This neuropterid fern is more typical of the late Pennsylvanian than of earlier Pennsylvanian fern fossils. Its occurrence in yellow, oxidized shale rather than that of a grey or dark color is also more typical of late Pennsylvanian plant occurrences as well as those of the Permian Period. Specimen at the right has the characteristic large terminal pinnule. Shale of the Pleasanton Group, Upper middle Pennsylvanian, St. Louis County, Missouri, both from localities now covered by urbanization. (Value range G).

Fig. 08-017. *Pecopteris* sp. A small specimen of this very common Pennsylvanian fern which constituted the foliage of a variety of ferns and seed ferns of Pennsylvanian coal swamps. Pleasanton Group, St. Louis County, Missouri. (Value range G).

Fig. 08-015. *Neuropteris flexuosa.* Two fronds from a paleokarst filling of the same age as the ferns of the two previously illustrated figures, but preserved as a compression. Pleasanton Group, Upper Middle Pennsylvanian. Shelbina, Missouri. (Value range F, single specimen).

Fig. 08-018. *Pecopteris* sp., frond. Part and counterpart. Pleasanton Group, St. Louis County, Mo.

Fig. 08-019. Left: *Aleothopteris* sp.; right: *Mixoneura* sp. With its large terminal leaflet the *Mixoneura* is particularly characteristic of the late Pennsylvanian, which is the age of these specimens. Monongohela Series, Upper Pennsylvanian, Nickolas County, West Virginia. (Value range G, single specimen).

Fig. 08-016. *Asterotheca* sp. Frond with part and counterpart from a creek in what was once the edge of the St. Louis metropolitan area, a locality now buried by urbanization, the fate of many fossil sites. Pleasanton Group, Middle Pennsylvanian, St. Louis County, Missouri. (Value range F).

Fig. 08-020. Another group of late Pennsylvanian fern fossils. Monongohela Series, Nickolas Co., West Virginia.

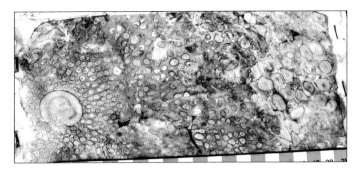

Fig. 08-023. *Psaronius* sp. This section through a large tree fern shows the tree's mantle made up of numerous roots surrounding the tree fern stem. Such a root mantle buttressed the plant, allowing it to stand upright in the soggy, wet environment of a coal swamp. This is an acetate peel from a coal ball. Coal balls are calcareous concretions which formed in the plant debris of what would eventually become a coal seam. They preserve with great detail the structure, including cellular structure, of the coal swamp plant. Coal by contrast, usually preserves little plant structure after the parent plant debris is converted to coal. Middle Pennsylvanian, Berryville, Illinois (Value range F).

Fig. 08-021. *Neuropteris* sp. Late Pennsylvanian ferns from siltstone beds of the Kansas City, Missouri, area. These ferns have some characteristics which align them with the plant life of the Permian Period. Winterset Formation, Upper Pennsylvanian. (Value range G).

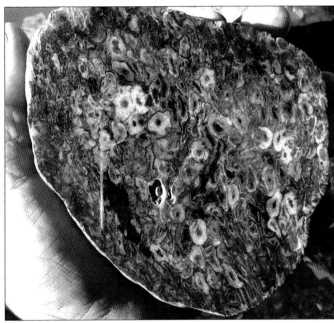

Fig. 08-024. *Psaronius* cf. *P. bliciklei*. This petrified portion of the root mantle of a tree fern was found as a glacial cobble in the St. Louis area. Petrified sections of *Psaronius* can be quite attractive. Some of the original plant material has been preserved in the same manner as in a coal ball, however in this case the preserving medium is quartz rather than the calcite of a coal ball. This specimen probably came from Middle Pennsylvanian strata of northern Illinois, carried southward by Pleistocene (ice age) glaciers. (Value range F).

Fig. 08-022. *Sphenopteris* sp. Pennsylvanian fern from sinkhole (paleokarst) deposits in the northern Ozarks. *Sphenopteris* is a fern which is associated with an upland flora that lived under conditions which were drier than the normal "coal swamp" flora more commonly seen. Such plants or their descendents proliferate in the dryer Permian Period of which this fern is suggestive. (Value range G).

Some of the tree ferns of the coal swamps bore seeds so they were not true ferns (which reproduce by spores). Determining which fossil foliage is from true ferns and which is from seed ferns is difficult as the various parts of a plant, seeds, foliage and wood, are generally found separately.

Fig. 08 025. Seeds of "tree ferns" (pteridosperms). Many of the so-called tree "ferns" of the late Paleozoic were not really ferns at all. Ferns don't reproduce by seeds; they reproduce by a more primitive method involving spores. Which coal swamp plants reproduced by seeds and which reproduced by spores is uncertain and unclear, as foliage and seeds are generally not found together. At the top is a calamite trunk section bearing the impression of a seed (*Pachytesta* sp., Brongniart). The heart-shaped seeds are *Trigonocarpon* sp., the most common seed of a pteridosperm. Pteridosperm is the (division) name given to the extinct seed "ferns" and is considered by paleobotanists to be the taxonomic equivalent of an animal phylum. Pteridosperms were not ferns at all, having evolved separately from ferns and not being related to them, but still having fern-like foliage, an example of convergent evolution, (Value range E for all).

Fig. 08-026. *Lyginopteris* sp. An acetate peel of a pteridosperm seed taken from a coal ball. Unlike compression fossils or seed casts such as those of the previous image, seeds preserved in coal balls show the internal structure of the seed in the same manner as can be observed in a modern seed. Middle Pennsylvanian, Berryville, Illinois. (Value range F).

Lycopods

Lycopods (lepidophytes), a division of the plant kingdom represented today by small, herbaceous plants such as lycopodium.

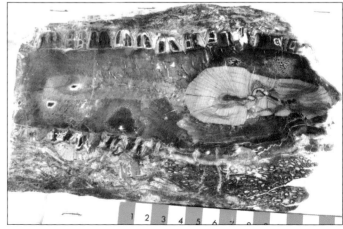

Fig. 08-028. *Stigmaria* sp. These are sandstone casts of the root of a lepidodendron. *Stigmaria* is found in clay layers which occur beneath coal seams (underclay) where they anchored the sizeable lepidodendron tree that grew in coal swamps; it is a fairly common Pennsylvanian plant fossil. Atoka Formation, Lower Pennsylvanian, Boston Mountains (southern Ozarks), Arkansas. (Value range F, single specimen).

Fig. 08-029. *Lepidophioios* sp. A compression fossil of a common Pennsylvanian lycopod. Lycopods and lepidodendrons belong to the Lycopodophyta, a major division (equivalent to phylum) of the plant kingdom. Crowberg Coal, Henry County, Missouri. (Value range F).

Opposite

Fig. 08-027. *Lepidodendron* sp. This coal ball peel is of a section of a lepidodendron or scale tree. Leaf scars on the surface of a lepidodendron are formed by the pattern produced from the attachment of elongate, needle-like leaves. Middle Pennsylvanian, Berryville, Illinois. (Value range F).

Fig. 08-030. Lepidodendron cone. This is a small lycopod or "scale tree," preserved in an ironstone nodule or concretion. Braidwood flora, Braidwood, Illinois. (Value range F)

Fig. 08-031. Left: *Lepidodendron* sp. A typical impression of the surface of a lepidodendron tree trunk. Right: *Stigmaria* sp., the root cast of a lepidodendron. Atoka Formation, Lower Pennsylvanian, Boston Mountains (southern Ozarks), Arkansas. (Value range F, either specimen).

Fig. 08-032. *Lepidodendron* sp. Bark or surface impression, Atoka Formation, Lower Pennsylvanian, Ben Hur, Arkansas. (Value range F).

Fig. 08-033. *Lepidodendron* sp. A distinct sandstone cast of the bark of the "scale tree." Sometimes impressions of this plant are thought to be the impressions of a large, prehistoric snake. Lower Pennsylvanian, Tennessee, Illinois. (Value range F).

Fig. 08-034. *Lepidodendron* sp. Distinct impression showing the characteristic branching of a lepidodendron tree. Fine sandstone has preserved these plants very well and a number of them from this locality have been distributed from Lower Pennsylvanian outcrops near Pella, Iowa. (Value range E).

Fig. 08-035. An illustration of Pennsylvanian outcrops on the Des Moines River from David D. Owen "Geology of the states of Iowa and Minnesota etc." 1852. The fossils of the previous image came from either these outcrops or ones near them near Pella, Iowa.

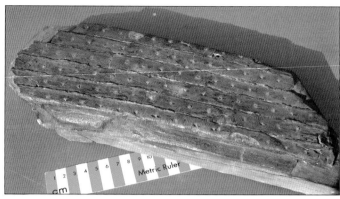

Fig. 08-036. *Sigillaria* sp. Surface impression of another type of distinctive Pennsylvanian scale tree. Middle Pennsylvanian, Hickory County, Missouri. (Value range F).

Fig. 08-037. *Sigillaria orbicularis,* Bronghiart. Ironstone concretion. Middle Pennsylvanian, Braidwood, Illinois. (Value range G).

Fig. 08-038. *Sigillaria* sp. Another coal age scale tree, this type having a leaf scar pattern different from that of *Lepidodendron*. Pleasanton Group sandstone, Middle Pennsylvanian, Grand River, Missouri. (Value range F).

Sphenophyllum

Fig. 08-039. *Sphenophyllum* is a common compression fossil in Pennsylvanian shales. It was a small, herbaceous plant that lived beneath the taller plants of the coal swamps. Middle Pennsylvanian, Henry County, Missouri.

Fig. 08-040. *Sphenophyllum* sp. Specimen at the right was collected in the late 19th century from coal pits mined at that time; the recovered fossil plants became the subject of contemporary paleobotanical works. Both specimens from Middle Pennsylvanian strata, Henry County, Missouri.

Arthrophyta

Arthrophyta is the plant division to which *Calamites* and *Annularia* belong. These were giant scouring or horsetail rushes some of which reached eighty feet in height.

Fig. 08-041. *Annularia* sp. The foliage of calamites. Calamites were a sort of gigantic scouring rush, having leaf whorls arranged in a regular pattern like this. Pleasanton Group, Mid Pennsylvanian, St. Louis County, Missouri. (Value range G).

148

Fig. 08-042. *Annularia* sp. A smaller set of calamite leaf whorls similar to and from the same locality as the previously illustrated specimen. (Value range F).

Fig. 08-043. Calamites and fern (*Pecopteris*). A calamite stem which looks similar to a horsetail or scouring rush. The scouring rush or *Equisitum* is a modern relative or descendent of calamites. Kinkaid Formation, Lower Pennsylvanian, Kinkaid Creek, southern Illinois. (Value range F).

Fig. 08-044. *Calamites* sp. Section of calamite wood from Pennsylvanian sea cliffs at Joggins, Nova Scotia. The Joggins sea cliffs, now a fossil preserve, was visited by Charles Lyell in the 1840s. He was puzzled by the similarity of these rocks (and their associated fossils) with those of the same age in Wales. Continental drift was still well in the future, but the strata at Joggins and similar strata of Wales were formed close together, later being separated by the opening of the Atlantic.

Fig. 08-046. *Calamites suckowi*, Brongniart. The terminal portion of a young plant of this distinctive arthrophyte. This is a cast of *Calamites* preserved in sandstone. Warrensburg-Moberly Sandstone, St. Louis County, Missouri. (Value range E).

Fig. 08-045. *Calamites suckowi*, Brongniart. A typical "reedy" impression of a calamites stem. This is a very common appearance of a calamites fossil. Middle Pennsylvanian (Upper Carboniferous), Joggins, Nova Scotia. (Value range F).

Fig. 08-047. *Calamites* sp. A coal ball peel of the cross section of a calamites stem which includes a considerable amount of woody material surrounding the pith (circular area) of the plant. Most calamite fossils are fillings of this pith with sediment such as sand, which then makes a pith cast. This particular *Calamites* had an abnormal amount of wood surrounding the pith. Middle Pennsylvanian, Berry, Illinois. (Value range G).

Cordaites

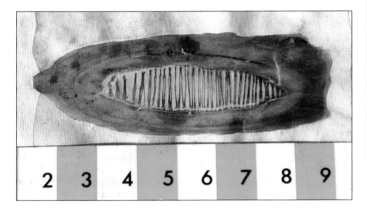

Fig. 08-048. *Cordaites* sp. An acetate peel of a section of the cordaites plant. *Cordaites* is considered by many paleobotanists as being an early conifer. Middle Pennsylvanian, Berryville, Illinois. (Value range G).

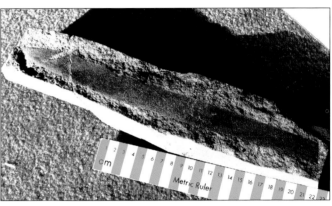

Fig. 08-050. *Cordaites* leaf compression. Long, strap-like leaves characterized this coal age plant that some paleobotanists consider may have been an early conifer. Warrensburg-Moberly Sandstone, Pleasanton Group, Middle Pennsylvanian, Grand River, Missouri. (Value range F).

Fig. 08-049. *Cordaites* sp. Top: cast of cordaites, which is given the form genus of *Artisia* sp.; bottom: *Cordaites* sp. from fossil drift wood mats of the Atoka Formation, Lower Pennsylvanian, Boston Mts., Arkansas.

Sponges

Sponges can occur locally in abundance in marine Pennsylvanian strata.

Fig. 08-051. Sponge. A late Paleozoic sponge sometimes found forming the nucleus of chert concretions, Middle Pennsylvanian, St. Louis County, Missouri. (Value range G)

150

Cnidarians

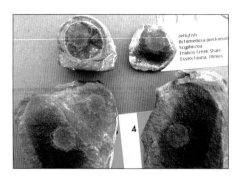

Fig. 08-052. *Octomedusa pieckorum*. Shown here are soft bodied impressions of small jellyfish from the Essex conservatat-lagerstatte of northern Illinois. Like the primarily non-marine Braidwood ironstone concretions, the fossils from these marine beds form a nucleus within the concretion. (Value range G).

Fig. 08-053. Button corals. These aptly named corals come from Lower Pennsylvanian strata of northeastern Oklahoma. (Value range H, single specimen).

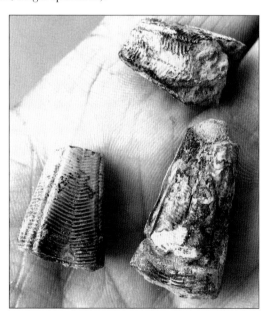

Fig. 08-054. *Conularia* sp. Conularids are exampled of problematic Paleozoic fossils which have been placed by various paleontologists, into different phyla, including mollusks, the annelids, and cnidaria. Some paleontologists also consider conularids to have been representatives of an extinct phylum. The blue color of the specimen on the left is from the original phosphate material of the animals exoskeleton. Upper Pennsylvanian, Platte County, Missouri. (Value range G, single specimen).

"Worms"

Fig. 08-055. *Rhaphidophorus hystrix*. Fossil polychaete worms are found in abundance in the Essex ironstone concretions. Many of the organisms found in these concretions have been "shoehorned" into the phylum annelida, a common practice in paleontology when elongate "worm-like" fossils are classified. The millipede designation of these Essex fossils has been well established from much rarer and clearer specimens. (Value range G, single specimen).

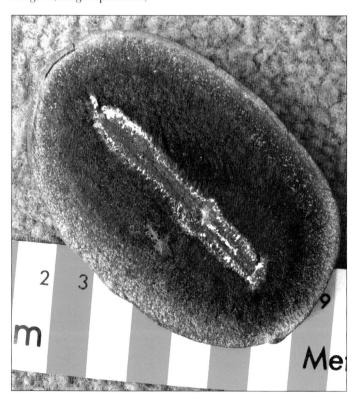

Fig. 08-056. *Didontogaster cordylina*. "Tummy tooth worm." Another fossil "worm" in ironstone nodules. Essex Biota, Pit 11, northern Illinois.

Brachiopods

Trace Fossils

Fig.08-057. *Echinoconchus* sp.(upper left), *Neospirifera* sp. (middle) and *Dictyoclostus* sp. (right). These are some of the larger and more frequently found Pennsylvanian brachiopods in marine strata of the U.S. midwest. Middle Pennsylvanian, St. Louis Co., Missouri (Value range F, all specimens).

Fig. 08-058. *Juresania* sp. A group of productid brachiopods from Pennsylvanian strata of the Brooks Range of northern Alaska. Mississippian and Pennsylvanian fossils of the Brooks Range can be almost identical to those of mid-continent North America. The strata yielding them in Alaska may be autochronous in that it was derived from somewhere else, as a large part of Alaska seems to be made of crustal rocks derived from other parts of the globe. Lisburne Formation, Atiken Gorge, northern Brooks Range, Alaska.

The tracks and trails made by (often unknown) organisms on or in sediments.

Fig. 08-059. *Helminthopsis* sp. This and the next three images are examples of trace fossils. Trace fossils represent the tracks, burrows, or trails of organisms (usually animals but not always) moving on the surface of sea floor sediment or burrowing within it. Trace fossils usually are given scientific names known as form genera, since the taxonomic position of its maker is often unclear or unknown. This trace fossil is representative of a type made by organisms living in relative deep water, where the animal is burrowing or moving through sea floor sediment. With this strategy the animal extracted organic matter from the sediment upon which it fed. Such deep sea trace fossils often have a geometric pattern that maximizes the efficiency of the burrowing pattern, so sediment that the animal has previously gone through is not crossed again. This strategy results in geometric patterns such as this spiral or other patterns of geometric interest. Such trace fossils can be relatively abundant in deep sea sediments. Deep sea sedimentary strata are known as flysch deposits, and often the only fossils found in them are trace fossils. Deep water facies of the Atoka Formation, Lower Pennsylvanian, Ouachita Mts., Arkansas. (Value range G).

Fig. 08-060. *Hormosiroidea* sp. Known as the "chain of beads" trace fossil. *Hormosiroidea* can be a fairly common trace fossil in Pennsylvanian sandstones. Middle Pennsylvanian, Fox River, Clark County, Missouri. (Value range G).

152

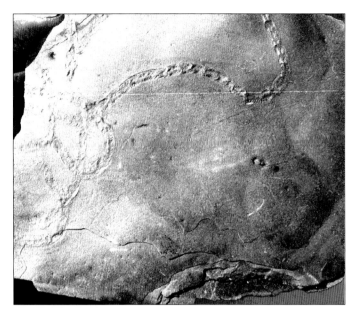

Fig. 08-061. This type of trace fossil is attributed to gastropods or some other mollusk moving slowly over sediment of the deep sea floor. When trace fossils occur, there is often a large number and variety of them occurring together. Atoka Formation, (deep sea facies, flysch), Ouachita Mountains, Arkansas. (Value range H).

Fig. 08-062. *Phycoides* sp.These are trace fossils believed to be the "feeding burrows" of worm-like animals which lived on what was the continental shelf of North America during the Paleozoic Era. This, like many trace fossils, is a cast or sediment filling of the burrow where the animal making it entered and left its sea floor burrow. Atoka Formation, ramp facies, Lower Pennsylvanian, Boston Mountains, Arkansas. (Value range F).

Blastoids

Blastoids are an extinct echinoderm class.

Fig. 08-063. *Pentremites brentwoodensis*. These are the youngest known blastoids from the northern hemisphere where the Brentwood Limestone, of lowermost Pennsylvanian age, yields them in northwest Arkansas and adjacent northeast Oklahoma. (Value range F for slab of specimens).

Fig. 08-064. *Pentremites brentwoodensis*. Individual blastoids are the norm for specimens like these of the genus *Pentremites*. *Pentremites* normally is considered as characteristic of strata of the Late Mississippian. This species is found in strata considered as early most Pennsylvanian age by stratigraphers (geologists who are involved in the specialty of determining the age of rock strata from fossils). Brentwood Formation, Early Pennsylvanian, Brentwood, Arkansas. (Value range G for single specimen).

Holothurians

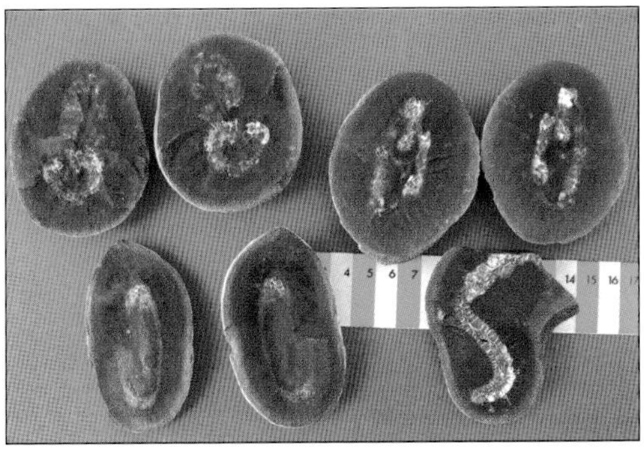

Fig. 08-065. *Achistrum* sp. Holothurians (sea cucumbers) are (or were) of local frequent occurrence in the Essex, Illinois, ironstone concretions. Usually, however, they are not very distinct. Exceptional and relative rare specimens have proved that these fossils are sea cucumbers, an echinoderm class that is otherwise poorly represented in the fossil record other than by the microscopic sclerites which are embedded in the animal's skin. Middle Pennsylvanian, Essex, Illinois. (Value range H for single specimen)

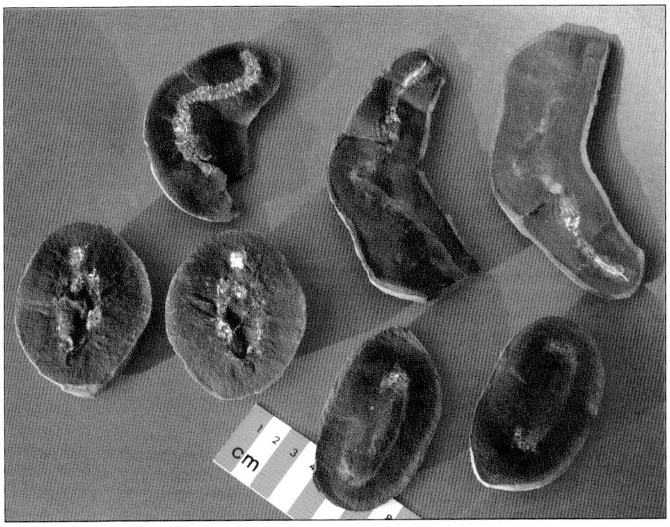

Fig. 08-066. *Achistrum* sp. A group of fossil holothurians preserved in ironstone concretions from the Essex locality. (Value G for a single specimen).

Crinoids

Very successful echinoderms in seas of the Pennsylvanian that periodically covered parts of continents.

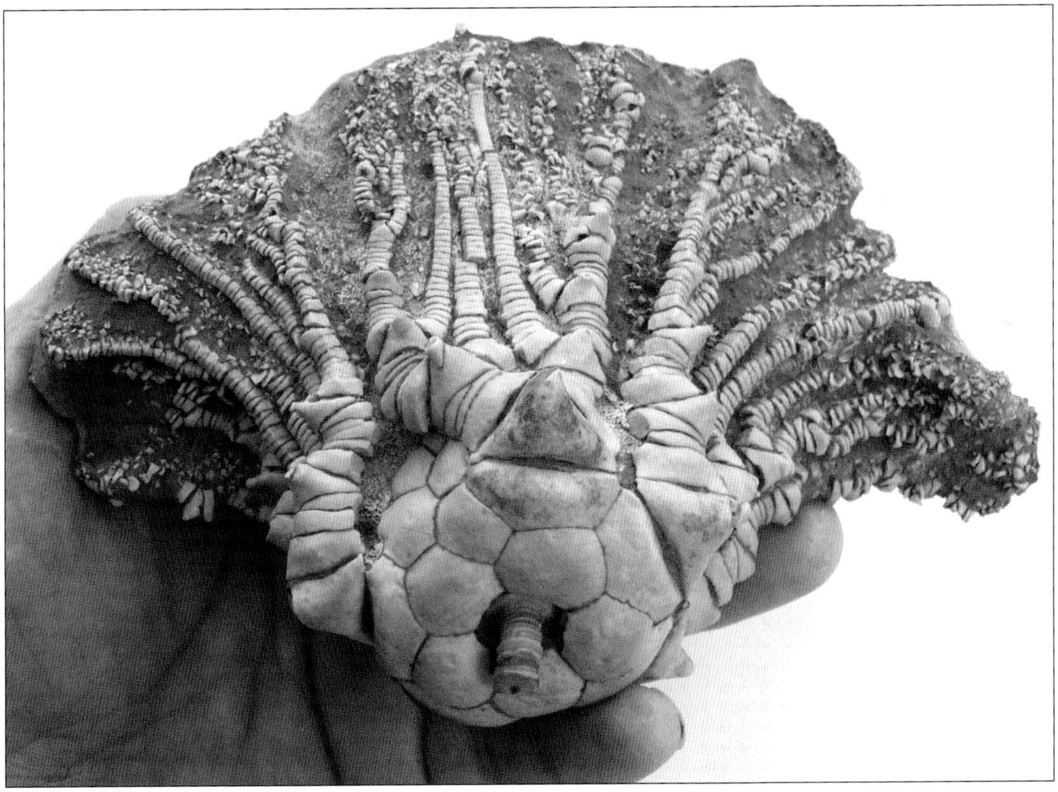

Fig. 08-067. *Metaperimestocrinus spiniferus*, Strimple. A large and spectacular Pennsylvanian crinoid with a pronounced anal sac umbrella. Holdenville Formation, Middle Pennsylvanian, Beggs, Oklahoma. (Value range D).

Fig. 08-068. *Metaperimesto-crinus spiniferus*, Strimple. Top view of the same specimen as above, showing the anal sac umbrella. The author of this crinoid, Harold Strimple, discovered and described large numbers of fossil crinoids in Oklahoma, Kansas, Missouri, and Illinois during the 1950s, 1960s and 1970s.

Fig. 08-069. *Delocrinus* sp. A relatively common type of crinoid found in Pennsylvanian strata of the U.S. midwest. Most frequently however, only the dorsal cups of the crinoid are found. Lane shale, Kansas City, Missouri. (Value range F).

Fig. 08-072. *Erisocrinus typus*. The typical occurrence for this attractive crinoid from the Pontiac or Ocoya, Illinois, crinoid locality. LaSalle Limestone, Upper Middle Pennsylvanian (Missourian). (Value range D).

Fig. 08-070. *Stellarocrinus* cf. *S. virgilensis*. A group of specimens of this attractive crinoid from the LaSalle Limestone near Pontiac, Illinois. Numerous fine and complete specimens have come from this fossil crinoid "garden" with *Stellarocrinus*, with their twenty arms bent inward, being the most common crinoid. (Value range for single specimen F).

Fig. 08-071. *Stellarocrinus* sp. Numerous complete crinoid crowns have come from shale beds overlying part of a thin limestone bed quarried in the 1960s and 1970s near Pontiac, Illinois. One of the most common crinoids in the fauna, it is this genus that has been widely distributed among collections. (Value range F for single crown).

155

Fig. 08-073. Right: *Onychocrinus* sp.; left: *Exocrinus* sp. Both from the LaSalle Limestone, Pontiac, Illinois crinoid locality. (Value range E).

Mollusks

Mollusks, presumably gastropods.

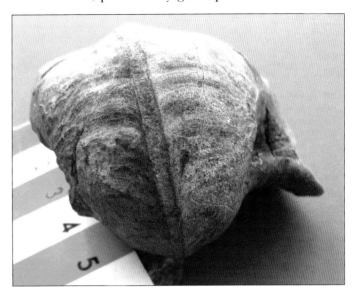

Fig. 08-074. Large Pennsylvanian crinoid stems. These large crinoid stems can be common fossils of the Pennsylvanian Period in North America, particularly from the middle part of the system in the U.S. midwest. They are never found with a calyx and it has been suggested that they may have had an unmineralized head or calyx. These specimens are from Middle Pennsylvanian Limestone, St. Louis County, Missouri. (Value range H for single specimen).

Fig. 08-075. *Bellerophont* sp. A planispiral gastropod characteristic of the Paleozoic Era. The bellerophonts represent peculiar planispiral mollusks; they are generally considered to be gastropods, but some paleontologists have considered them to be a type of monoplacophoran. What is peculiar about the family Bellerophonticea is their bilateral symmetry. They are coiled in a plane like a coiled nautaloid cephalopod, which they somewhat resemble, but, unlike a cephalopod, they lack chambers or septa and of course had different soft parts. Specimen from Middle Pennsylvanian of Colorado. *Courtesy of Kenneth Brill.* (Value range G).

Fig. 08-076. *Bellerophont* sp. A group of bellerophonts from Middle Pennsylvanian strata that locally can contain relatively large numbers of these interesting gastropods. Marmanton Group, Middle Pennsylvanian, St. Louis County, Missouri, (Value range F, single specimen).

156

Cephalopods

Cephalopods were successful mollusks in Pennsylvanian seas.

Fig. 08-077. *Temnochellus (Metacoceras) sangamonense*. A portion of this peculiar nautaloid bears large, rounded spines. This cephalopod is rarely found complete, but rather specimens are usually fragmented. It may have been crushed by and its soft parts eaten by shark-like fishes. Middle Pennsylvanian, St. Louis County, Missouri. (Value range E).

"Tullymonsters"

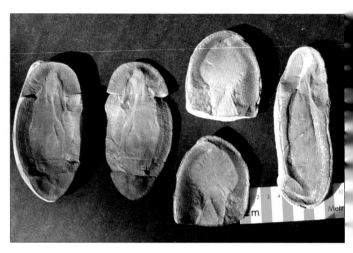

Fig. 08-079. *Tullimonstrum gregarum*. The tulley monster, a problematic fossil believed by some paleontologists to be a type of mollusk, or even a representative of an extinct phylum. They are found in the ironstone concretions of northern Illinois. Middle Pennsylvanian, Essex fauna, pit 11, northern Illinois. (Value range G for single good partial like shown here. Complete specimens can go for much more).

Fig. 08-078. *Titanoceras ponderosum*. A large coiled nautaloid from Upper Pennsylvanian strata of the Kansas City area. A number of these as well as other large cephalopods have come from dimension stone quarries in the Kansas City area. Winterset Limestone, Upper Pennsylvanian, Kansas City, Missouri (Value range E).

Trilobites

Trilobites and trilobite trackways.

Fig. 08-080. *Ditomophyge* sp. *Ditomophyge* is one of the more frequently found Pennsylvanian trilobites, a group which becomes greatly reduced in number and diversity after the Devonian. Marmanton Group, Middle Pennsylvanian, Ferguson, Missouri. (Value range F).

Fig. 08-081. *Ameura* sp. A trilobite of the late Paleozoic. Trilobites become less diversified and abundant after the Devonian Period. Pennsylvanian and Permian trilobites are relative rare, and their diversity is even less than with trilobites of the Mississippian Period. (Value range F)

Fig. 08-083. These are fossil track ways. Probably they were made by trilobitomorphs or what are referred to as soft-bodied trilobites. They appear to have been locally quite abundant during the Paleozoic Era and their trackways are found in areas where sediments were deposited very rapidly. (Value range F, for individual specimen).

Fig. 08-084. This is a short trilobiteomorph track. It is actually a cast of the track. The original track was made in very soft mud which was then covered by a layer of fine sand. The sand filled the track making this hyporelief cast. Atoka Formation, deep sea facies, Ouachita Mountains, Arkansas.

Fig. 08-082. *Diplichnites* sp. This trace fossil is the track way of a trilobitomorph, a trilobite-like animal which lacked an exoskeleton. Trilobitomorph trackways can occur in Paleozoic fine sediments of deep sea origin known as flysch. One of the best known and documented trilobitomorphs is *Mariella splendens* from the Middle Cambrian Burgess Shale. In the Burgess shale actual soft part morphology of the trilobiteomorph can be seen, but usually only track ways like this are found. Note the radiating impressions around the maker's feet. These are the impressions of the animals gills which were located at the end of each jointed leg. Trilobitomorphs are of course arthropods like trilobites; they were a sort of soft bodied trilobite. Atoka Formation, Lower Pennsylvanian, Ben Hur, Arkansas. (Value range E, higher than most trilobitomorph trackways because of the gill impressions).

158

Other Arthropods

Fig. 08-085. *Crossopodia* sp. This zipper-like trace fossil is attributed to some type of arthropod living in a relatively deep sea environment. Atoka Formation, Boston Mountains, Ozone, Arkansas, Atoka Formation, ramp facies. Courtesy of Mike Fix (Value range E).

Fig. 08-086. *Euproops danae*. A fossil xiphosurian; a relative to the horseshoe crab. These are found associated with plants of the coal swamps. They are associated with the coal flora such as represented by the Braidwood, Illinois, concretions. (Value range E).

Fig. 08-087. *Kallidecthes richardsoni*. Shrimp from the Essex (pit 11) locality, which has preserved many soft bodied animals. It is considered as one of the late Paleozoic paleontologic "windows." (Contrast enhanced).

Fig. 08-088. *Kallidecthes* sp. Single shrimp specimen. Essex iron stone nodules, Pit 11, Essex, Illinois. (Value range F).

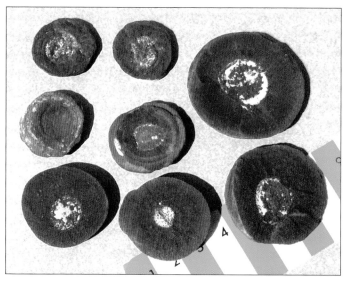

Fig. 08-089. *Cyclus americanus*. A peculiar crustacean with a rounded body, a fossil usually associated with the sediments above Illinois coal beds. Essex fauna. (Value range G, single specimen).

Fig. 08-090. *Belotelson magister*. A fossil lobster-like arthropod associated with both the Essex (pit 11) fauna and the Braidwood flora. (Value range E).

Fig. 08-091. Insect wing. Wings are the commonly found parts of fossil insects. This insect wing, from a cockroach, is preserved in an ironstone nodule of the Illinois basin, coal fields. Terre Haute, Indiana. (Value range E).

Sharks and Shark-like Fishes

Spines and teeth of sharks and shark-like fishes.

Fig. 08-093. *Listracanthus* sp. A spine that covered the surface of a peculiar shark-like fish of the Pennsylvanian of the U.S. midwest. *Listracanthus* is usually associated with black, organic-rich shale that lies above a widespread mineable coal seam in Indiana, Illinois, and Missouri. (Value range G, single specimen).

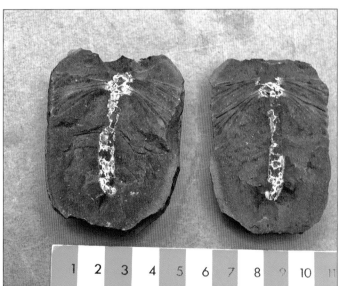

Fig. 08-092. Dragonfly. Insect flight in the Pennsylvanian was the earliest attempt at flight in the animal kingdom. Such Middle Pennsylvanian dragonflies are the oldest known of this "first in flight" insect. Middle Pennsylvanian, Terre Haute, Indiana. (Value range D, rare).

Fig. 08-094. Slab covered with *Listracanthus* spines. Middle Pennsylvanian, Moberly, Missouri.

160

Fig. 08-095. *Petrodus* sp. Dermal tubercles of sharks. These small, cone shaped fossils can be locally abundant in Pennsylvanian strata. They were originally embedded in the skin of sharks that inhabited shallow marine waters, which were often preceded in time by a coal swamp. (Value range F, for group).

Fig. 08-096. *Petalodus* sp. Crushing teeth of shark-like fish or bradyodonts. Middle Pennsylvanian, Ferguson, Missouri.

Fig. 08-097. Shark coprolites, preserved as impressions in ironstone concretions or nodules. Impressions of spiral coprolites (fossil excreta) can be fairly common in Pennsylvanian ironstone concretions such as this. (Value range G).

Fig. 08-098. *Edestus heinrichi*. A number of these "jaw-like" sections of spines of peculiar Paleozoic shark-like fishes have come from the roof shale of coal mines of southern Illinois, particularly from underground mines near Sparta, Illinois. Middle Pennsylvanian. (Value Range D).

Fig. 08-099. *Edestus* spine on left with illustrations of a similar specimen from a late 19th Century publication on Pennsylvanian fish. Illinois Geological Survey, Vol. IV, 1870.

Coelacanths

Coelacanths are a type of crossopterygian or lobe finned fish.

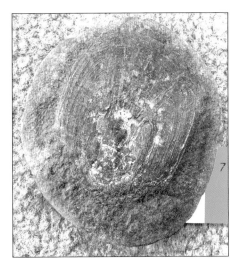

Fig. 08-103. Coelacanth scale. This is a scale of a large fish known as a Coelacanth. These are found in the braidwood concretions where these possible ancestors to amphibians ventured into the coal swamps. Braidwood, Illinois, Francis Creek Formation. (Value range G).

Fig. 08-100. *Edestus* sp. A variant on this shark spine found in Pennsylvanian age marine strata. These are often associated with the small dermal tubercles placed in the genus *Petrodus*. Both of these remains may be from the same animal, a shark-like fish that lived in shallow seas which periodically covered the continents of the northern hemisphere. The cyclic deposits of marine and non-marine sediments, characteristic of the Pennsylvanian, are called cyclothems. Middle Pennsylvanian, St. Louis Co., Mo.

Fig. 08-104. Fossil Coelacanth fish scales along with an informative and readable little book on the finding of living coelacanths in the Indian Ocean. Used book fairs where this book was found, can be a source for inexpensive and unexpected interesting works on fossils like this.

Fig. 08-101. *Edestus* sp. A relatively small form of this distinctive spine of a shark-like Pennsylvanian fish. These are often associated with the small dermal tubercles known as *Petrodus* (see above). Ferguson, Missouri (see above). (Value range G, single specimen).

Fig. 08-102. *Orocanthus* sp. A relatively large dorsal spine of a shark from a thin limestone bed associated with channel-coal in Upper Carboniferous (Pennsylvanian) strata of England. Birdiehouse Limestone, England. (Value range D).

Amphibians

Amphibian fossils (either tracks or body fossils) are desirable.

Fig. 08-105. Amphibian tracks. Tracks and track ways of tetrapods (amphibians) appear with some abundance locally for the first time in strata of the Pennsylvanian Period. Earlier tetrapod tracks are known, but they are quite rare. Such tracks and trackways record the first appearance of land vertebrates, which were the embolomere amphibians. Hartshorne Formation, Middle Pennsylvanian, northeastern Oklahoma. (Value range E).

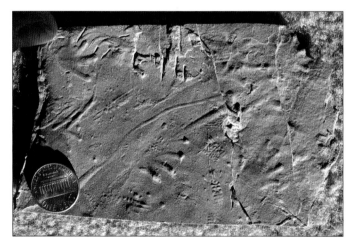

Fig. 08-106. Amphibian tracks. Bedding surfaces of Pennsylvanian strata can have track ways or tracks of these early amphibians. The tracks and track ways are more common than actual skeletal remains which are quite rare this early in the geologic record. Middle Pennsylvanian, Black Warrior Basin, northern Alabama. (Value range E).

Bibliography

Andrews, Henry N. *Ancient Plants and the World They Lived In.* Comstock Publishing Associates, Ithaca, New York, 1947.

Janssen, Raymond E. *Leaves and Stems from Fossil Forests.* Illinois State Museum, Popular Science Series, Vol. 1. Springfield Illinois, 1979.

Langford, George, 1963. *The Wilmington Coal Fauna* and *Additions to the Wilmington Coal Flora from a Pennsylvanian Deposit in Will County, Illinois.* Esconi Associates, 1963.

Owen, David D. *Report of the Geological Survey of Wisconsin, Iowa, and Minnesota and Incidentally of a Portion of Nebraska Territory.* Lippencott, Grambo and Co., Philidelphia, 1852.

Shabica, Charles W., and Andrew A. Hay, editors. *Richardson's Guide to the Fossil Fauna of Mazon Creek.* Northeastern Illinois University, 1995.

White, David. *Fossil Flora of the Lower Coal Measures of Missouri.* U S Geological Survey Monograph, 37, 1899.

Wittry, Jack. *The Mazon Creek Fossil Flora.* Esconi Associates, Downers Grove, Illinois, 2006.

Chapter 9
The Permian Period

The Last Period of the Paleozoic Era

The last period of the Paleozoic Era, the end of which marked what is considered to be the world's greatest extinction event. Sedimentary rocks of the Permian are often colorful, their various shades of red, orange, and variegated colors (purple and grey) reflecting the highly oxidized iron of what are appropriately called red beds. As a consequence of this world wide oxidization, it has been suggested that the oxygen level of the earth's atmosphere may have been somewhat higher than the 21% of today. Such a high oxygen level would promote these colorful, oxidized sediments as well as produce fierce forest fires of which there is some evidence in the form of thin charcoal layers.

The Permian was the time of Pangea, the so-called "supercontinent," a single landmass on the earth. The end of the Permian is referred to as the "terminal Paleozoic extinction event" with the existence of Pangea seemingly having had something to do with this event.

Plants

Permian plants are often different from the plants of the Pennsylvanian Period.

Fig. 09-01. *Pecopteris* sp. A Permian example of this ubiquitous Northern Hemisphere late Paleozoic "fern." Rotliegendes Series, Pfalz, Bei Maiuz, Germany.

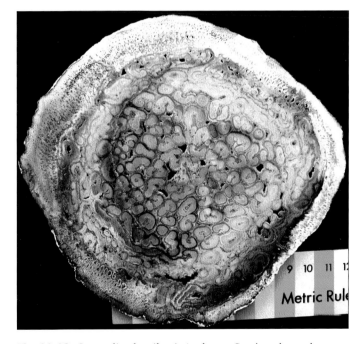

Fig. 09-02. Fern. Permian ferns and fern-like foliage are often more delicate and less robust than these Pennsylvanian ancestors. Wellington Formation, Upper Permian, Perry, Noble Co., Oklahoma. (Value range G).

Fig. 09-03. *Osmundites brazilensis* Andrews. Section through a colorful petrified trunk of an Osmundid tree fern from Permian strata of Brazil. Polished slices of this plant have been widely distributed through the fossil market. The osmundid ferns are a common living family and some representatives of the family make up tree ferns of the modern tropics. The earliest osmundid ferns appear in the Permian and representatives of the family are also found in the Mesozoic. The Pennsylvanian tree fern *Psaronius* is similar in structure to *Osmundites*, but is considered by paleobotanists as not related to the osmundids. *Psaronius*, like *Osmundites* had a buttress of roots allowing it to live under very swampy environments. (Value range F).

164

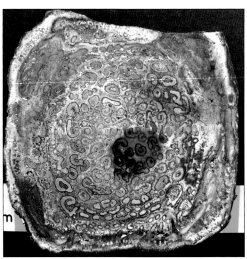

Fig. 09-04. *Osmundites brazilensis* Andrews. These specimens are labeled as Permian in age, however, Andrews (1961), the author of the species, states "the exact age of this fossil is not known but it is probably not older than late Mesozoic." The petrified tree fern logs, from which these specimens were sliced are associated with Gondwanaland sediments, which are Permian in age. Permian sediments and associated fossils are often found in red beds that are colorful like these sections of tree ferns. (Value range F).

Fig. 09-05. *Osmundites brazilensis* Andrews. A particularly colorful slab of the trunk of this petrified tree fern from Tocantins State, town of Filadelfia, Brazil. Large quantities of this petrified tree fern, as well as other petrified wood, are gathered in the vicinity of the Brazilian town of Filadelfia, where surface weathering brings out the color and pattern in the wood. (Value range F).

Fig. 09-06. *Glossopteris sp.* These distinctive fossil plants are found in Permian strata associated with Gondwanaland; the supercontinent that formed at the end of the Paleozoic Era and whose fossils are characteristic of the southern hemisphere. The distinctive series of strata containing these fossils in the southern hemisphere led early 20th century earth scientists to propose continental drift and its break up of the "supercontinent" of Gondwanaland. The *Glossopteris* flora is widespread on the fragments of Gondwanaland which include Australia, South Africa, South America, India, and Antarctica.

Fig. 09-06. *Glossopteris sp.* These distinctive fossil plants are found in Permian strata associated with Gondwanaland; the supercontinent that formed at the end of the Paleozoic Era and whose fossils are characteristic of the southern hemisphere. The distinctive series of strata containing these fossils in the southern hemisphere led early 20th century earth scientists to propose continental drift and its break up of the "supercontinent" of Gondwanaland. The *Glossopteris* flora is widespread on the fragments of Gondwanaland which include Australia, South Africa, South America, India, and Antarctica.

Fig. 09-08. A group of *Glossopteris* slabs. New South Wales, Australia.

Protists and Sponges

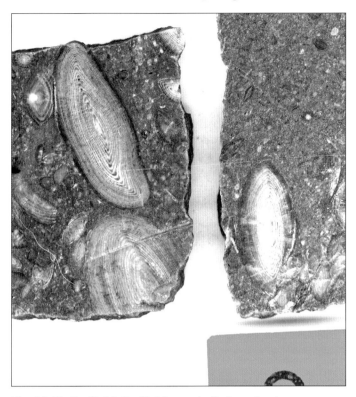

Fig. 09-09. Fusilinid. Fusilinids are shelled protists known as Foraminifera. This is quite large for a Protist and fusilinids underwent extensive speciation in the Pennsylvanian and Permian periods. They are reliable biostratigraphic markers in Pennsylvanian and Permian limestones as they changed rapidly over short periods of geologic time and also can be very abundant in Permian and Pennsylvanian limestones. Middle Permian, Northern France.

Fig. 09-10. This sponge is associated with the interior of chert nodules or concretions that occur in the Upper Permian Kaibab Limestone of northern Arizona. Similar fossil sponges are sometimes found associated with chert nodules in Pennsylvanian strata. (Value range F).

Brachiopods

Brachiopods are lower invertebrates which became minor elements in marine life fter the Permian extinction event.

Fig. 09-11. *Derbyoides* sp. A large brachiopod typical of the Pennsylvanian and Permian periods. The Permian is the last period of geologic time where brachiopods are abundant, diverse, and widespread. Most brachiopod genera of the Paleozoic Era go extinct at the end of the Permian Period. Brachiopods are really a Paleozoic type fossil! Fort Riley Limestone member of Barnston Formation, Chase County, Kansas. (Value range H).

Fig. 09-12. *Composita* sp. A brachiopod genus common in Pennsylvanian an Permian limestone and marine shale. Brachiopods are shelled invertebrates but they are not mollusks, They are in their own phylum, the phylum brachiopoda which in today's oceans make up a minor element of shallow water, benthonic (bottom dwelling) life. Florena Shale member, Beattle Formation, Dexter Kansas. (Value range, single specimen H).

Echinoderms

Echinoderms make up this group with some types, like the blastoids, going extinct at the end of the period in the "Terminal Permian Extinction Event."

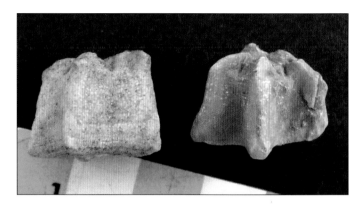

Fig. 09-15. *Timorblastus* sp. Blastoids of this type represent the last occurrence of this extinct class of Echinoderms, they being one of the many Paleozoic life forms which went extinct at the Permian extinction event. Blastoids, however, are almost unknown in the preceding Pennsylvanian Period except for a species of the genus *Pentremites* which occurs in Lowermost Pennsylvanian strata of Arkansas and Oklahoma. *Timorblastus* is found only in the Permian of the Southern Hemisphere where it is associated with Permian sediments of Gondwanaland. (Value range F or G, single specimen).

Fig. 09-16. A group of *Timorblastus* from the Island of Timor. (top view).

Fig. 09-17. The same group of blastoids as in the previous picture but in side view. Courtesy of Larry Osterberger.

Fig. 09-13. *Prorichyofenia* sp. This is a reef building brachiopod that contributed to reefs of the Permian seas of west Texas and New Mexico. This genus and a number of related genera superficially resemble corals more than they do a brachiopod, An example of parallel evolution where a group of organisms (brachiopods in this case) unrelated to corals, evolved a "life style" similar to that of corals, resulting in an organism superficially resembling a coral when it adapted the coral "mode of life," that of a reef building organism. Word Limestone, Glass Mountains north of Marathon, Texas. (Value range G).

Fig. 09-14. *Boxtonia* sp. A large brachiopod belonging to the Paleozoic brachiopod family known as the productids. Early Permian, Junction City, Kansas. (Value range H, single specimen).

Fig. 09-18. *Deltoblastus batheri*. Another distinct (and terminal) blastoid from the Permian of Timor, Indonesia.

Fig. 09-19. Permian crinoids from Timor. Left: undetermined crown; right: *Timorechinus* sp. Value range G, single specimen of *Timorechinus*, F for specimen on left.

Fig. 09-21. Another group of *Jimbocrinus bastocki* from the Permian of Western Australia.

Fig. 09-22. Another group of these peculiar Permian crinoids. Note the coiled tegumen on specimen at the lower left.

Fig. 09-20. *Jimbocrinus bastocki*. These peculiar and distinctive crinoids come from outcrops along the Gascoyne River valley in western Australia. A number of these came onto the fossil market in 1995. Introduction of these distinctive crinoids into the fossil market expanded the diversity of crinoids available from parts of the world other than North America and Europe. Unfortunately Australian regulations and export authorities stopped export of additional material which occurs fairly extensively in western Australia and is part of the same Permian strata that yields the Timor fossils. (Value Range E for an average specimen like these).

Fig. 09-23. Three calyces of *Jimbocrinus basstocki*. Permian of Western Australia. (Value range G, single specimen).

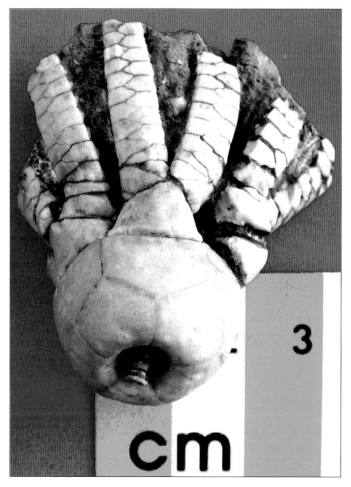

Fig. 09-24. *Delocrinus* sp. A type of crinoid generally more typical of the Pennsylvanian than of the Permian. Specimens of this crinoid occur infrequently in the lower Permian of Kansas and Oklahoma. Lower Permian, Johnson City, Kansas. (Value range F).

Mollusks

Mollusks were a very successful phylum in the Paleozoic. Many types went extinct either during or at the end of the Period. These included these coiled nautaloids of a type which had been around since the Lower Ordovician. The pearly nautilus type of nautaloid, a cephalopod which appeared in the Mesozoic, survived to the present but it is quite different from these archaic nautaloids,

Fig. 09-25. *Protodentalium* sp. These are fossil scaphapods (also known as tusk shells). Scaphapods belong to a class of mollusks which became common in the Pennsylvanian and Permian periods. There are elongate, scaphapod-like fossils in the Ordovician, however these have features which suggest they are not scaphapods. Scaphapods form part of the molluscan fauna of modern oceans. Upper Permian, Roswell, New Mexico.

Fig. 09-26. Two specimens of a coiled nautaloid from the Upper Permian Kaibab Limestone of the Colorado Plateau, northern Arizona. Such cephalopods are characteristic Paleozoic nautaloid cephalopods and are similar with those found earlier in the Pennsylvanian. The end of the Permian marked the extinction of most nautaloids along with most other invertebrate life which had dominated Paleozoic seas for millions of years. The Kaibab Limestone forms the rim of the Grand Canyon and caps much of the Colorado Plateau of northern Arizona. The Kaibab Limestone is overlain by Mesozoic strata (Moenkopi Formation), the interval between these two formations marking the boundary between the Paleozoic and Mesozoic eras as well as the Paleozoic extinction event. (Value range E for single specimen).

Arthropods

These are all arthropods. Trilobites went extinct at the end of the Permian, but were small and inconspicuous in the Permian. Insects on the other hand diversified and proliferated to the diversity which exists today where they are our neighbors.

Fig. 09-27. *Ditomopyge decurtata.* Permian trilobites are similar to those of the Pennsylvanian, but rarer. Lower Permian, Beattie Formation, Florena Shale Member, Cowley Co., Kansas. (Value range G).

Fig. 09-28. Group of insect wings from late Permian lake deposits (fresh water limestone). Insects diversified in the Permian; they are generally fragmentary remains found in fresh water limestone layers that formed in relatively localized lakes. Wellington Formation, Perry, Noble County, Oklahoma.

Fig. 09-29. Dragonfly wing, Permian lake deposits. This is a wing of a large dragonfly. Some of the largest dragonfly's ever to exist are found in Permian strata usually associated with localized fresh water lakes. (Value range F).

Fish

A selection of Permian "fish."

Fig. 09-30. *Lawnia* sp. These fossil fish are known as paleoniscids or ray-finned fish. Paleoniscids were widespread in brackish water environments that were particularly widespread during the Permian in many parts of the world. Lueters Formation, Seymour, Texas. (Slabs with this many Paleozoic fish on them are rare).

Fig. 09-31. *Palaeoniscus freieslebeni* Blainville. Paleoniscids have a characteristic "fish-like" (homocercal tail) which is bilaterally symmetrical as seen on this specimen. Marl Slate, Upper Permian, Old Quarrington Hill, County Durham, England.

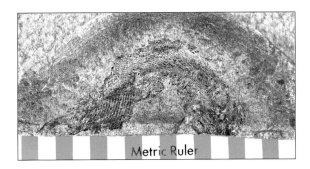

Fig. 09-32. *Palaeoniscum freieslebeni* Blainville. This paleoniscid fish is preserved in part of a concretion. Upper Permian, Old Quarrington Hill, County Durham, England. (Value range F).

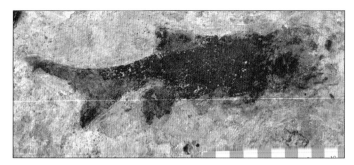

Fig. 09-33. *Paramblypterus duvernoyi*. A paleoniscid fish preserved as a carbon film in fresh water lake sediments similar to those of England and Scotland. This Permian deposit of Permian Lake sediments in Germany has yielded a number of fish as well as amphibians. Lower Rotliegendes, Odernheimy, Pfalz, Germany. (Value range E).

Fig. 09-34. Another specimen of a Palaeoniscid fish from Upper Permian strata of Saxony, Germany.

Fig. 09-35. *Amblipterus macropterus*. A paleoniscoid fish preserved in ironstone. Permian strata often are reddish in color (red beds), however strata which yield fossil fish and amphibians are usually dark colored and high in organic material. This fossil is preserved in oxidized, iron rich sediments of an ancient lake bed. Most Permian red beds are notably baren of fossils except for fossil tracks and trackways. Otovice, Bohemia, Czechoslovakia. (Value range E).

Fig. 09-36. *Pleurocanthus*. These are the teeth of fresh water sharks known as pleurocanths. They first appear in the Pennsylvanian but become particularly abundant during the Permian where their teeth can locally be common fossils. (Value range F, for group).

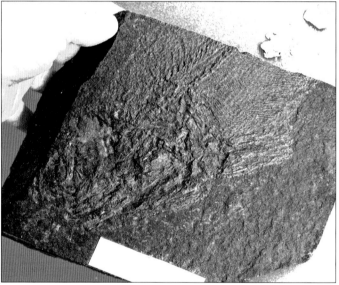

Fig. 09-37. Coelacanth or lobe-finned fish from the Permian Kupferschiefer (copper slates) of middle Rhineish Germany. The fish is preserved by being coated with a thin, yellowish layer of chalcopyrite, a copper mineral. Many sedimentary rocks of Permian age contain cpper, sometimes in quantities that make such strata a source of copper (copper ore). The coelacanth is a type of fish suspected to have been an ancestor of tetrapod amphibians which led to reptiles. Upper Permian, Kupferschiefer, Zechstein Group, eastern Germany. (Value range E).

Amphibians

Here is a delight of Permian amphibians or the tracks made by them. Tetrapods (four-legged land vertebrates) did well in the Permian even with its locally arid conditions.

Fig. 09-38. *Branchiosaurus (Paranblypterus) Paramblystomus* sp. Branchiosaurs are late Paleozoic freshwater amphibians. Branchiosaurs of this genus are the Paleozoic amphibians normally seen on the fossil market. Most of these come from Permian brackish or fresh water deposits of Germany (Pfalz) or from Czechosolovakia.

Fig. 09-39. *Discosaurus saurisscus*. A branchiosaur. Lower Rotliegendes, Brno (Brunn) Czechosolovakia.

Fig. 09-41. Paleozoic fossil amphibians are generally rare fossils although there are some exceptions. On the right are illustrated specimens from the same work as mentioned above. Specimens illustrated from the book are from Late Pennsylvanian or Early Permian lake deposits at Linton, Ohio. On the left are Permian branchiosaur amphibians from a locality which has produced many similar specimens. Oderheim, Pfalz, Germany.

Fig. 09-40. Fossil amphibians are found associated with sediments deposited in fresh or brackish water. Such sediment was deposited in lakes which were abundant during the Permian. The reconstruction of early amphibians is from a major collectable work on Paleozoic fossil tetrapods.

Fig. 09-42. Close-up of specimen at bottom left of previous picture. Oderheim, Pfalz, Germany. (Value range F).

Fig. 09-43. Labyrinthodont teeth. These amphibian teeth show a radiating pattern of the infolding of the tooth enamel from which the name of this group of amphibians is derived, The tooth has a labyrinth-like structure. A similar tooth structure is seen on Crossopterygian (lobe-finned) fishes. This similarity is one of the strong evolutionary connections between these two groups of vertebrates. Lower Permian, Seymour, Texas. (Value range G).

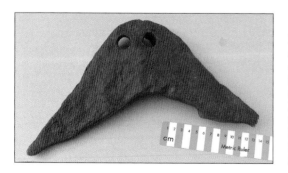

Fig. 09-44. *Diplocalus* skull-cast. A cast of the bony skull of a specialized Permian amphibian. Permian of Texas. (Value range G, for cast).

Fig. 09-45. Stegocephalian skull (cast). The stegocephalians were large amphibians with massive, bony skulls. Fragments of bone head amphibians can be fairly common fossils in some fresh water Permian lake and stream deposits. Complete skulls, however, are rare and usually require painstaking piecing together of the usually highly fragmented fossils. Once a specimen has been reconstructed, casts such as this can be made relatively easily. Clear Fork Formation, Lower Permian, Texas.

Fig. 09-46. Amphibian trackways, Coconino Sandstone. The Coconino Sandstone represent a series of fossil sand dune deposits of an alternately wet and dry climate. These trackways were made by small amphibians walking up the surface of sand dunes. The Cococino sandstone is quarried in northern Arizona for dimension stone that is used extensively in the American Southwest and these trackways can turn up on the bedding planes. The Coconino Sandstone also forms the upper wall of the Grand Canyon (below the Kiabab Limestone which forms the canyons rim rock). Amphibian trackways similar to these can be seen on some of the sandstone slabs along the canyon's trails. (Value range F).

Fig. 09-47. Amphibian tracks, Coconino Sandstone. Another slab of Coconino Sandstone with amphibian tracks from a northern Arizona dimension stone quarry. (Value range F).

Fig. 09-48. Amphibian trackways. Casts of the tracks of small amphibians cover this colorful siltstone slab from the Upper Permian near Abilene, Texas. (Value range F).

Fig. 09-49. Bedding planes on large slabs of Coconino Sandstone from along the San Juan River of Utah. Such trackways can be locally abundant on bedding plane surfaces of this wind and dune deposited sandstone of Middle Permian age.

Reptiles

These are reptiles! Most Permian reptiles went extinct at the end of the period, however there are glimpses of the "Age of Reptiles," including *Edaphisaurus* and *Dimetridon* seen here; models of which commonly accompany prehistoric animal toy models.

Fig. 09-50. Large talus boulder along the San Juan River, Utah, with trackways of what may be a reptile. Coconino Sandstone, Middle Permian.

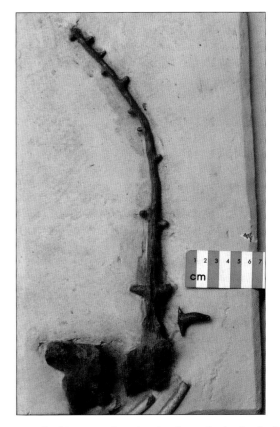

Fig. 09-51. *Edaphisaurus* spine. A spine from the back of a "sail back" reptile similar to *Dimetridon*. *Dimetridon* and *Edaphasaurus* are early reptiles, but they are not dinosaurs, even thought *Dimetridon* models are often seen with dinosaur models. Lower Permian, Clear Fork Formation, Seymour, Texas.

Fig. 09-52. *Dimetridon* skull cast. This is a plaster cast of a restored skull of this Permian Pelycosaur. It was one of the sail-back reptiles of the Permian. Note the dentition of *Dimetridon*, an obvious carnivor. A number of these casts have come from Jone's Fossil Farm, Worthington, Minnesota. (Value range E).

Fig. 09-55. *Dimetridon* bronze models. *Dimetridon* and Edaphisaurus often come with models of dinosaurs however neither one of these is a dinosaur. They are pelycosaurs and unlike all dinosaurs, they are Paleozoic, **not** Mesozoic, in age. These attractive (but pricy) bronze models were available in museum gift shops before the 1990s public "surge" of interest in dinosaurs. (Value range F, single specimen).

Fig. 09-53. *Edaphisaurus* spine showing lateral projections coming off of the spine. *Edaphisaurus* and *Dimetrodon* are both pelycosaurs. *Edaphisaurus* has a dentition type suggestive that it was omnivorous. The spine of *Dimetrodon* lacks the side bars of *Edaphisaurus*. Clear Fork Formation, Lower Permian, Seymour, Texas. (Value range E).

Fig. 09-56. *Dimetridon* sp. These are plastic models from the recent "era" of the "Jurassic Park" dinosaur craze. They are cheap, plastic kids toys which, it should again be emphasized are **not** dinosaurs but pelycosaurs, anatomically an entirely distinct, extinct group of Paleozoic reptiles.

Fig. 09-54. *Dimetridon* dorsal spine fragment. The fin spines of this pelycosaur lacks the lateral side bars of Edaphisaurus. An Edaphisaurus "side bar" is to the left of the *Dimetridon* spine. These Texas Permian fossils are uncommonly seen, as they occur on large cattle ranches that discourage collecting. Clear Fork Formation, Lower Permian, Seymour, Texas. (Value range F).

Fig. 09-58. Some of the small bones of stem reptiles etched from Permian breccia like that above. Permian paleokarsts, Arbuckle Mountains, Lawton, Oklahoma. (Value range G).

Fig. 09-61. *Sphenodon* skull-cast. A plaster cast of the skull of a stem reptile. From Jones Fossil Farm, Worthington, Minnesota. (Value range H).

Fig. 09-59. Another assemblage of small bones of stem reptiles (probably *Capttorhiuns* sp.) etched from Lower Permian bone breccia. Lawton, Oklahoma. (Value range F).

Opposite

Fig. 09-57. Boulder of stem reptile bone breccia. This jumble of fossil bones comes from ancient sink holes (paleokarsts) which formed at the edge of the Arbuckle Mountains of Oklahoma. Such sinkholes were filled with Permian sediments chuck full of small bones of early reptiles and amphibians. These fossil bones are normally black, but on exposure they weather to a bluish tint. Such bones can be etched from the breccia with acetic acid and complete skeletons can be painstakingly reconstructed from them. Black components in the rock are the fossil bones.

Fig. 09-60. *Sphenodon* sp. A skeleton of one of the stem reptiles reconstructed from bones etched from breccia that come from Permian paleokarsts of Oklahoma. This specimen was exhibited at the Field Museum of Natural History of Chicago. Courtesy of Field Museum.

Fig. 09-62. *Mesosaurus brazilensis*. An example of a superb, articulated fossil reptile. *Mesosaurus* and *Glossopteris* are the two fossils most associated with the concept of continental drift, particularly with regards to Gondwanaland. The southern part of the supercontinent of Pangea which existed during the Permian Period. These articulated Permian reptiles entered the fossil market in quantity during the mid 1980s. They come from Permian fresh water limestone associated with Gondwanaland sedimentary rock. In these fresh water limestones they can locally occur in some abundance. Articulated vertebrates of most types are rare fossils (except for some fossil fish) and the availability of these excellent specimens enabled institutions and individuals to have an example of a complete reptile. Unfortunately, the Brazilian government stopped the exportation of these fossils a few years ago as they are considered to be national treasures. If a fossil is locally abundant it serves best to regard it as an educational resource and to disseminate it accordingly. Such dispersal of the numerous specimens of *Mesosaurus* was a win-win situation for both the locals who quarry the fossils and also for individuals and institutions who desire such items. They may be "national treasures," but they are abundant enough to go to all of those who desire and wish to purchase them. Locking up fossil localities, which after all are in layers of sedimentary rock that usually extend over a considerable area, doesn't really make sense, unless it is to keep an item scarce and exclusive. The archeological model (the reason for often restricting paleontological resources) applied to paleontology is rarely valid, as the occurrence of the resources of these two fields is quite different. (Value range C).

Bibliography

Gould, Steven J. "The Great Dying" in *Ever Since Darwin. Reflections in Natural History,* W. W. Norton Co., New York, NY, 1977

Glossary

Arkose. A type of sandstone composed of fragments of granite (mechanically weathered granite). Arkose is characteristic of continental or cratonic parts of the earth that have undergone rifting, where the arkosic sediment filled in an opening rift.

autochronous terrain. A part of a continent made up of rock and rock strata that appears foreign to the region in which it occurs. Fossils found in areas of autochronous strata will be different and "out of place" from those of native strata. Autochronous terrain is explained by plate tectonics as rock strata originating from some other part of the planet then being transported by sea floor spreading over considerable distances and added or welded to the place where it is currently found.

body fossils. Consisting of shells, tests, bones, teeth or even impressions of soft parts of an animal. A body fossil contrasts with a trace fossil, which is a burrow, track, or other evidence of movement which can be associated with the behavior of an organism.

Cambrian radiation event. The event which resulted in a proliferation of body plans in organisms some 540 million years ago at what marks the beginning of the Cambrian Period of geologic time. Fossils (both trace and body fossils) become obvious and relatively abundant after this time.

Canadian shield. That part of the stable continental interior (craton) of North America which exposes at the surface very ancient rocks of Precambrian age. Most of "the shield" lies in Canada, however parts of it are in Minnesota, Wisconsin, upper Michigan, and New York.

coal balls and coal ball peels. Coal balls are concretions found in coal beds which can preserve in great detail, parts of the plants which formed the coal. Coal itself rarely contains well preserved fossils but associated coal balls embed woody tissue with calcareous material and preserve plant structure. A coal ball peel is usually made by coating a sliced and etched coal ball with an acetate sheet softened with a solvent. The sheet picks up a layer of plant material that becomes embedded in the film, and in many ways resembles a photographic negative.

division. In botany, a major subdivision of the plant kingdom. A division in botany is the equal to a phylum in the animal kingdom.

flysch. Sedimentary rocks usually devoid of normal marine fossils such as shells but containing many tracks and trails (trace fossils) of deep sea origin and usually deposited rapidly so that such trace fossils can be preserved. Flysch deposits represent the sediments of the open ocean or a deep sea trench. Flysch deposits can also form parts of mountain ranges where the sediments have been transported (by sea floor spreading) and added to what was, at one time, a margin of a continent, an island arc, or part of a continent.

greywacke. A type of dirty, often iron-rich silty sandstone usually associated with regions involved with tectonic activity, such as subduction zones. Greywacke can be deposited in either marine or non-marine environments and is indicative of tectonic activity at the time of deposition. It is rare or lacking in the shallow water marine sediments deposited on a stable continent or craton.

ICZY or International Commission on Zoological Nomenclature. A group of zoological taxonomists who meet periodically (usually every ten years) to decide issues having to do with the validation of scientific zoological names, including those of fossil animals and animal trace fossils. A similar commission exists for botany.

miogeosyncline. A sedimentary depositional environment consisting of relatively thick, shaley or slaty (clastic) sediments deposited in relatively deep marine waters. This usually is indicative of sediments deposited near the edge of a continent whose continental crust has been pulled downward by the process of subduction. Miogeosynclines of the past can form the rocks of mountain ranges and such rocks can be fossiliferous. The Atlas Mountains of North Africa is an example of a site of a former miogeosyncline.

paleokarst. An ancient, sediment-filled sinkhole (karst). Sinkholes offer a way to preserve such younger sediments and their fossils. These fossils will be younger and "out of place" when compared to those of the surrounding older rocks, which will usually be limestone or dolomite as these are the rocks which commonly form sinkholes.

phylum. The highest level of classification in the animal kingdom. In the Linnaean system of classification the levels descend from phylum to class, to order, to family, genus, and, finally, species.

primary literature. Scientific publications which describe and illustrate for the first time an organism previously unknown to science. Such scientific literature is always referenced by experts seeking to place fossils in their correct group.

protists. Single-celled organisms that reproduce by sharing genetic material with another cell.

178

rift zone. An opening (or divergence) of the earth's felsic or continental crust that can initiate the formation of a new ocean basin. The Red Sea Rift in the Middle East is probably the best known modern example of a rift zone. Rift zones can fill with sediment from rivers draining into them and such sediments can contain fresh water animal fossils, such as fish, trackways of land animals, such as dinosaurs, as well as land plants.

sea floor spreading. A process whereby new ocean crust is formed moving previously formed crust. Sediments deposited on such an ocean floor move with the ocean floor, which "spreads" to accommodate the creation of new crust at a mid-ocean ridge. This provides a mechanism to move sediments of deep sea origin and their fossils and pile them against a continent while oceanic crust (mafic material) sinks back into the earth's mantle. This process of moving deep sea sediments and piling them against a continental margin has made ancient deep sea sediments and their fossils accessible.

steinkern. The interior mold of a fossil animal. A steinkern can form in the interior of a bivalve or other mollusk or in a crinoid as well as other shelled animals when a cavity is filled with sediment. On the removal of the shell or other hard part (by solution when the hard part is buried in sediment), a natural mold of the interior of the fossil is formed; such a mold is a steinkern (German: stone center).

Subduction zone. A place in the earth's crust where the ocean floor (mafic or oceanic crust) is dragged back into the earth's mantle producing a deep sea trench where the downward movement of oceanic crust takes place. This process can "scrape" sea floor sediments and pile them at the edge of a trench. This adjacent landmass may be an Island Arc or a continent. Such sediments from the deep sea are referred to as eugeosyncline sediments and can include flysch deposits.

taxonomy. The science of classification of some group of things, usually animals or plants. This is conventionally done using what is called the Linnaean System of nomenclature, which is a hierarchal series of different levels of relationships, with phylum and division being the highest ranks in the animal and plant kingdoms respectively.

tectonic regions. In plate tectonics these are regions of the earth associated with specific types of movement of parts of the earth (plates). A plate can crack forming a rift zone and the pieces can then spread away from each other forming a linear ocean like the present day Red Sea.

vendozoans. A group of puzzling, radially or bisymmetrical fossil impressions of a life form that apparently had a tough, leathery composition. Vendozoans are found in rocks of the late Precambrian (Neoproterozoic) and represent one of the first clear occurrences of complex life forms. Some paleontologists consider vendozoans to represent an extinct high taxonomic level of life viz. a Kingdom. Also known as Ediacarian organisms from the Ediacarian Hills of northeast Australia where the organisms were first discovered. Some Cambrian fossils such as *Camptostroma* and Archeocythids have been suggested as being types of vendozoans, otherwise vendozoans appear to have become extinct at the end of the Proterozoic Era.

Index

O

Olenellid trilobites, 31
oncolites, 16
Ozarkian and Canadian, 40-55

P

paleontology and archeology, 5
Plants
    arthrophytes, 147-149
    cordaites, 149
    ferns, 140-143
    Glossopteris, 164
    lycopods, 114-115, 145-146
    psylophytes, 96
    sea weed, 114
    tree ferns, 97, 139, 143-144. 163-164
plate tectonics, 19-11
Polypia, 69

R

Receptaculites, 62-63, 82
rift zones, 12

S

sedimentary rocks, 5
Sponges, 18, 41, 63, 82, 98, 149, 165
stromatolites, 17, 41

T

Tentaculites, 103
trace fossils, 11, 21, 38, 83-84, 151, 152, 157
Trilobites, 31-37, 56-60. 72-79, 90-93, 197-198, 134, 157, 169
Tullymonsters, 156

V

Vertebrates
    agnatha (ostracoderms), 81, 95
    lobe-finned fish, 161, 170
    paleoniscids, 137, 169-170
    placoderms, 108
    reptiles, 173-176
    shark-like fishes (bradyodonts), 135-136, 159-161
    sharks, 170

W

"worms", 21-22, 64, 150